Assessment in Mathematics Education Contexts

This book aims to provide theoretical discussions of assessment development and implementation in mathematics education contexts, as well as to offer readers discussions of assessment related to instruction and affective areas, such as attitudes and beliefs. By providing readers with theoretical implications of assessment creation, revision, and implementation, this book demonstrates how validation studies have the potential to advance the field of mathematics education. Including chapters addressing a variety of established and budding areas within assessment and evaluation in mathematics education contexts, this book brings fundamental issues together with new areas of application.

Jonathan D. Bostic is an associate professor of mathematics education at Bowling Green State University, Bowling Green, OH, USA.

Erin E. Krupa is an assistant professor of mathematics education at North Carolina State University, Raleigh, NC, USA.

Jeffrey C. Shih is a professor of mathematics education at the University of Nevada Las Vegas, Las Vegas, NV, USA.

Routledge Research in Education

This series aims to present the latest research from right across the field of education. It is not confined to any particular area or school of thought and seeks to provide coverage of a broad range of topics, theories and issues from around the world.

Recent titles in the series include

Personal Narratives of Black Educational Leaders
Pathways to Academic Success
Edited by Robert T. Palmer, Mykia Olive Cadet, Kofi LeNiles, and Joycelyn L. Hughes

Exploring Digital Technologies for Art-Based Special Education
Models and Methods for the Inclusive K-12 Classroom
Edited by Rick L Garner

Subjectivities, Identities, and Education after Neoliberalism
Rising from the Rubble
Abraham P. DeLeon

Arts-Based Teaching and Learning in the Literacy Classroom
Cultivating a Critical Aesthetic Practice
Jessica Whitelaw

Quantitative Measures of Mathematical Knowledge
Researching Instruments and Perspectives
Edited by Jonathan D. Bostic, Erin E. Krupa, and Jeffrey C. Shih

Assessment in Mathematics Education Contexts
Theoretical Frameworks and New Directions
Edited by Jonathan D. Bostic, Erin E. Krupa, and Jeffrey C. Shih

For a complete list of titles in this series, please visit www.routledge.com/Routledge-Research-in-Education/book-series/SE0393

Assessment in Mathematics Education Contexts

Theoretical Frameworks and New Directions

Edited by Jonathan D. Bostic,
Erin E. Krupa, and Jeffrey C. Shih

LONDON AND NEW YORK

First published 2019
by Routledge

2 Park Square, Milton Park, Abingdon, Oxfordshire OX14 4RN
52 Vanderbilt Avenue, New York, NY 10017

Routledge is an imprint of the Taylor & Francis Group, an informa business

First issued in paperback 2020

Library of Congress Cataloguing-in-Publication Data
A catalog record for this book has been requested

ISBN: 978-1-138-59871-3 (hbk)
ISBN: 978-0-367-67076-4 (pbk)

Typeset in Sabon
by Apex CoVantage, LLC

Contents

List of Contributors   vii
Acknowledgments   xii

1   Introduction: Aims and Scope for *Assessments in Mathematics
    Education Contexts: Theoretical Frameworks and
    New Directions*   1
    JONATHAN D. BOSTIC, ERIN E. KRUPA, AND JEFFREY C. SHIH

2   Developing an Instrument With Validity in Mind   12
    MATTHEW R. LAVERY, CINDY JONG, ERIN E. KRUPA, AND
    JONATHAN D. BOSTIC

3   Measure Validation as a Research Methodology for
    Mathematics Education   40
    ERIK JACOBSON AND REBECCA BOROWSKI

4   Gathering Validity Evidence Using the BEAR Assessment
    System (BAS): A Mathematics Assessment Perspective   63
    MARK WILSON AND DIANE B. WILMOT

5   Validity Arguments for Instruments That Measure
    Mathematics Teaching Practices: Comparing the M-Scan
    and IPL-M   90
    TEMPLE A. WALKOWIAK, ELIZABETH L. ADAMS, AND
    ROBERT Q. BERRY

6   Design and Validation Arguments for the Student Survey
    of Motivational Attitudes towards Statistics (S-SOMAS)
    Instrument   120
    DOUGLAS WHITAKER, ALANA UNFRIED, AND MARJORIE BOND

7 **Measuring Self-Efficacy to Teach Statistics in Grades 6–12 Mathematics Teachers** 147
LEIGH M. HARRELL-WILLIAMS, JENNIFER N. LOVETT,
LAWRENCE M. LESSER, HOLLYLYNNE S. LEE, REBECCA L. PIERCE,
TERI J. MURPHY, AND M. ALEJANDRA SORTO

8 **Measurement and Validity in the Context of Mathematics Coaches** 172
KRISTIN E. HARBOUR, STEFANIE D. LIVERS, AND MARGRET
A. HJALMARSON

Index 196

Contributors

Elizabeth L. Adams is a STEM evaluation researcher with Research in Mathematics Education at Southern Methodist University. Her research focuses on quantitative program evaluation of STEM interventions and measuring mathematics and science classroom practices using tools such as observations and instructional logs. Previously, she taught fourth grade and was an evaluation consultant with Wake County Public School System in Raleigh, North Carolina.

Robert Q. Berry III is President of the National Council of Teachers of Mathematics and is a Professor in the Curry School of Education and Human Development at the University of Virginia. Equity issues are central to Berry's research efforts focused on understanding Black children's mathematics experiences, measuring standards-based mathematics teaching practices, and unpacking equitable mathematics teaching and learning. His most recent work has focused on using qualitative meta-synthesis in mathematics education research.

Marjorie Bond is a full professor at Monmouth College, Monmouth, IL, USA. She has worked in various areas of statistics education but since 2006 has concentrated on affective constructs, specifically, in the area of attitudes towards statistics and students' perceptions of statistics. He has been a co-leader of various research groups on affective construct. She mentors new researchers and keeps pushing the research forward.

Rebecca Borowski researches students' developing quantitative reasoning as they work with linear representations of quantity such as number lines. Her work operates at the intersection of three big ideas – how students learn to count and conceptualize of discrete quantities, how they learn to measure, and how they construct notions of continuous quantity. She is also interested in the differences between teachers' and students' conceptions of linear representations of quantity.

Jonathan D. Bostic's primary area of scholarship is exploring issues and trends within the context of quantitative assessment and evaluation in mathematics education. This area includes both development and uses of quantitative instruments for intended outcomes. Secondarily, he investigates ways to enhance instructional contexts to better support teaching and learning, especially learners' mathematical proficiency. This area includes work within inservice and preservice teacher education and professional development settings, as well as K-20 students' learning experiences and outcomes.

Kristin E. Harbour's scholarly agenda includes: (1) support systems for general and special education teachers' mathematics content knowledge and pedagogical content knowledge to advance their inclusive mathematics instruction, (2) interventions and innovative instructional practices that lead to improved mathematical understanding for all learners, and (3) pre-service teacher preparation with a focus on authentic experiences to navigate the complexities of the teaching and learning of elementary mathematics.

Leigh M. Harrell-Williams's interests include statistics and mathematics education, using elements of social cognitive theory to explore students and teachers' beliefs, attitudes, and motivation as predictors of behavior and achievement. Her primary research in this area focuses on K-12 mathematics teacher efficacy to teach statistics. She is also interested in instrument development using a Rasch measurement theory framework.

Margret A. Hjalmarson's work has focused on mathematics education and engineering education. In particular, he researches mathematics teacher leadership development, STEM faculty development and the use of design-based research methods in STEM Education projects.

Erik Jacobson studies how mathematics teachers develop the knowledge and beliefs that matter most for student learning, including beliefs related to equity and inclusion. He develops novel assessment and learning tools for teacher education, and then uses these tools for research aimed to better understand the complexity of teacher education and to find new ways to improve teaching.

Cindy Jong's research applies the complementary constructs of identity and affect (i.e., attitudes, beliefs, and dispositions) to understand how elementary teachers position themselves in relation to contexts and experiences that might influence their classroom practices. She also uses the framework of professional noticing to extend research into classroom practices and has an interest in assessing and measuring the aforementioned constructs.

Erin E. Krupa strives to make quality mathematics education more equitable to all students by researching the design, dissemination, and effectiveness of curricular materials and innovative professional development for mathematics educators and studying teachers' implementation of instructional materials. Erin's research pays close attention to the opportunity to learn students are provided within a classroom and how teachers can increase this index for all students, regardless of demographics.

Matthew R. Lavery studies the development, validation, use, and improvement of educational assessments and psychological instruments. He investigates the analysis of data to support valid inferences and decisions that inform reflection and improve the outcomes of professional practice, programs, and policy in the areas of teaching, learning, and leadership.

Hollylynne S. Lee's expertise is in teaching and learning of probability, statistics, and data science. She is an expert on the design of technology tools to facilitate students' learning and preparing teachers to use technology effectively. She directs the Hub for Innovation and Research in Statistics Education with a strong team of researchers, graduate students, and technological designers. Her work includes developing and implementing online courses and microcredentials to prepare teachers to teach statistics and data science.

Lawrence M. Lesser informed by experiences as a state agency statistician, textbook author, journal editor, high school teacher, university teaching center director, and PI on NSF grants, Lesser's mathematics/ statistics education research seeks to develop and assess ways to make mathematics/statistics more intuitive, engaging, and meaningful. Foci are: teacher education/knowledge, multiple representations, misconceptions, intuition/counterintuitiveness, equity (including English learners, culture/ethnomathematics, ethics, diversity), and engagement (including educational song). His 100+ papers include articles in top-tier journals (e.g., Statistics Education Research Journal).

Stefanie D. Livers's research agenda focuses on three areas: teacher preparation, teacher support, and equitable teaching practices. Specifically, my interests lie in teacher candidates' and practicing teachers' beliefs, knowledge, and critical perspective, teacher preparation program evaluation, professional development and coaching as tools for teacher support and growth, and meaningful mathematics opportunities for diverse students.

Jennifer N. Lovett's research is centered around the development of pre-service teachers' mathematical and statistical knowledge for teaching with a focus on utilizing dynamic technology tools to support students' learning.

Gabriel Matney researches authenticity in the learning and teaching of mathematics across world cultures with a purview on both classroom and professional development contexts. He seeks to develop instruments with sufficient validity evidence to be used in research on and about mathematics learning and teaching.

Teri J. Murphy's research interests are in discipline-based education research (DBER) in undergraduate science, technology, engineering, and mathematics (STEM); recruitment, retention, and graduation of undergraduates; diversity, equity, access, and inclusion; preparation and development of graduate students as instructors of undergraduates; and statistics education.

Rebecca L. Pierce, Associate Professor of Mathematical Sciences, has authored or co-authored professional articles and books related to statistics, statistics education, mathematics, and gifted education. From 1996 to 1998, she directed a National Science Foundation Grant under the EHR Model Projects for Women and Girls Program. From 1999 to 2010, she was a co-author, curriculum developer, and trainer for several Jacob K. Javits' grants, the only federal funding available for gifted programs.

Jeffrey C. Shih is a professor at the University of Nevada, Las Vegas. Currently, he is interested in doctoral preparation in mathematics education and the development and use of quantitative measures.

M. Alejandra Sorto's research focuses on the preparation of teachers in the area of Statistics, the impact of professional development, and comparative studies in Latin-America and Africa. In particular, she is interested in developing instruments to measure content knowledge for teaching, teaching quality, and analyzing its effect on student achievement. She has worked with governments of Chile, Peru, Dominican Republic, Honduras, and Guatemala to help improve the preparation of teachers in mathematics and develop educational standards. In 2011, the National Science Foundation (NSF) awarded her a CAREER research grant to investigate the Mathematics instruction of English language learners in the state of Texas.

Alana Unfried is Assistant Professor of Statistics at California State University, Monterey Bay. Her research interests focus on undergraduate statistics education, including the creation of co-requisite courses to support introductory statistics students, and the development of instruments assessing student and instructor attitudes towards statistics. She also enjoys statistical consulting work and collaborating with STEM education faculty on assessing K-12 student attitudes toward and interest in STEM.

Temple A. Walkowiak is Associate Professor of Mathematics Education at NC State University. Her research focuses on the measurement of mathematics instructional quality and on teacher knowledge, beliefs, and practices in K-5 mathematics. Her most recent work has included a longitudinal study of elementary teacher development from pre-service to the early years of teaching and a study examining teacher knowledge and instructional practices in fractions and decimals for both novice and veteran teachers.

Douglas Whitaker is Assistant Professor at Mount Saint Vincent University in Halifax, Canada. His research interests include affective constructs related to learning and teaching statistics and the development of high-quality instruments for research purposes. Dr. Whitaker's current work focuses on the relationship between instruments and theoretical frameworks in statistics education, collecting appropriate validity evidence to support the development and revision of measures, and improving the statistical preparation of teachers.

When **Diane B. Wilmot** graduated from Northwestern University's School of Education and Social Policy in 1996, she envisioned leading a school's efforts to teach children through innovative curriculum design and implementation. In 1998, she started as a high school math teacher in NY and then continued to teach math at Los Altos High School. Dr. Wilmot left the classroom to receive her Ph.D. in Education from UC Berkeley's Graduate School of Education. Her research mainly focused on measuring and evaluating students' procedural knowledge and conceptual understanding in mathematics. As a past administrator for the county and Palo Alto Unified School District, she has applied her research in classrooms and supported school leaders and teachers in their efforts to focus on academic growth and make learning more visible in order to improve student learning.

Mark Wilson is a professor of education at the University of California, Berkeley, and also at the University of Melbourne. He teaches courses on measurement in the social sciences, multidimensional measurement and applied statistics. His research focuses on the development of sound frameworks for measurement, new statistical models, instruments to measure new constructs, and on the philosophy of measurement.

Acknowledgments

Reviewers

We wish to sincerely thank several reviewers who provided feedback on chapters: Stephanie Casey, Theodore Chao, Chris Engledowl, Daria Gerasimova, Maryann Huey, Kristin Lesseig, and Jeremy Zelkowski

National Science Foundation

This book is a product generated from a National Science Foundation–funded conference held April 2–3, 2017 (Validity Evidence for Measurement in Mathematics Education; #1644314, 1644321). Any ideas expressed in this book by authors are those of the individuals and do not reflect the views of the National Science Foundation.

Personal

The editors wish to dedicate this book to their families.

Jonathan—To Brynn, Matthew, and Josephine.
Erin—To Marianne, Carolina, and Tyler.
Jeffrey—To Meg, Abby, and Penelope.

1 Introduction

Aims and Scope for *Assessments in Mathematics Education Contexts: Theoretical Frameworks and New Directions*

Jonathan D. Bostic, Erin E. Krupa, and Jeffrey C. Shih

The aim of this edited book is twofold. First, it provides readers with a deeper understanding of validity, validity evidence and arguments, and the validation process; differing approaches to the validation process; and authentic examples of ways to convey validation arguments for assessments. The book also includes chapters addressing a variety of established and budding areas within assessment and evaluation in mathematics education contexts. Thus, its audience is intended to include anyone who conducts assessment and measurement work in mathematics education contexts.

The American Educational Research Association, American Psychological Association, and National Council on Measurement Education ([AERA, APA, & NCME], 2014) provide clear guidelines regarding measurement validity and reliability in the *Standards for Educational and Psychological Measurement in Education* (hereafter referred to as the *Standards*). Sufficient evidence for five sources must be shared related to validity: (1) content evidence, (2) evidence for relationship to other variables, (3) evidence from internal structure, (4) evidence from response processes, and (5) evidence from consequences of testing (AERA et al., 2014). Unfortunately, "evidence of instrument validity and reliability is woefully lacking" (Ziebarth, Fonger, & Kratky, 2014, p. 115) in the mathematics education literature. Worse yet, evidence related to the validity of quantitative assessments and measures has not necessarily been conceptualized or defined consistently in the research literature (Lissitz & Samuelsen, 2007; Mislevy, 2007). In the last 10 years, there has been a concerted response by some working within the mathematics education research space to address this omission; these authors present validity arguments and purpose statements and discussion of them in an effort to better support research with quantitative measures (see AERA et al., 2014; Kane, 2016, 2001; Pellegrino, DiBello, & Goldman, 2016; Schilling & Hill, 2007; Wilson, 2005).

Readership and Intentions

This book has the potential to be used in a variety of contexts and by a wide audience. It should be noted for readers that this book has a complement— *Quantitative Measures in Mathematical Knowledge: Researching Instruments and Perspectives* (Bostic, Krupa, & Shih, 2019)—which focuses on validity arguments related to mathematics content measures. We, the editors, purposefully wanted two separate books because assessments and measures are different in nature (AERA et al., 2014). A test or measure is "an evaluative device or procedure in which a systematic sample of a test taker's behavior in a specified domain is obtained and scored using a standardized process" (AERA et al., 2014, p. 224), whereas an instrument may or may not involve correct or incorrect answers but rather Likert scales, indicators, or other criteria (Krupa, Carney, & Bostic, 2019). Assessment is broader than an instrument or measure (AERA et al., 2014). The *Standards* (AERA et al., 2014) define assessment as "any systematic method of obtaining information, used to draw inferences about characteristics of people, objects, or programs; a systematic process to measure or evaluate the characteristics or performance of individuals, programs, or other entities, for purposes of drawing inferences" (p. 216) or concomitantly as "a process that integrates test information with information from other sources (e.g., information from other tests, inventories, and interviews; or the individual's social, education, employment, health, or psychosocial history)" (p. 2). Hence, assessment includes a broader and more inclusive notion than measure or test. To that end, this book focuses on *assessments* for use in mathematics education contexts.

Second, we, the editors, intend for this book to be widely accessible to readers, including academic professionals, graduate students, industry experts working within the educational space, as well as practitioners. We are particularly excited about two areas of potential readership. The first is that this book may serve those working within the mathematics education space as well as related fields such as learning sciences, cognitive science, psychometrics, research assessment and evaluation, policy, special education, and other fields. Given a wide readership, a goal from this book and its complement is to encourage synergistic work across diverse scholars, which results in knowledge that has strong intellectual merit and broader impact. As editors, we are also excited to support those who are new to assessment. Reviews of assessment work within mathematics education research (e.g., Beckman, Cook, & Mandrekar, 2005; Bostic, Krupa, Shih, & Carney, 2019; Bostic, Lesseig, Sherman, & Boston, in press; Boston, Bostic, Lesseig, & Sherman, 2015; Hill & Shih, 2009) indicate that validity rarely comes up in peer-reviewed mathematics education scholarship. Moreover, when it does, the validity evidence centers on content and internal structure evidence. Connecting validity to these two sources of evidence is woefully lacking in building a robust validity argument that supports valid interpretations and uses and does not adhere to current *Standards* (AERA et al., 2014), much less prior *Standards* (AERA, APA, & NCME, 1999). To that end, we intend for graduate students, postdoctoral scholars, practitioners, and seasoned assessment veterans to reflect on their current assessment development practices and how this is linked with modern *Standards* (AERA et al., 2014).

A chapter by Lavery, Jong, Krupa, and Bostic (2019, this volume) provides readers with a broad overview of the validation process and ways to gather evidence. Examples are shared to instantiate the subprocesses during validity evidence collection, specific to each validity source. With greater volume and quality of work being done in this area, we anticipate seeing more manuscripts being submitted to peer-reviewed journals. In turn, this may cause some journal editors to reevaluate what is and is not appropriate for potentially publishable manuscripts, specifically the value of validity arguments related to assessments used within mathematics education research and the interpretations drawn from those assessments.

Validity and Validation

It is expected that assessments, broadly speaking, develop a purpose statement for users, and convey a validity argument. A purpose statement provides important information about an assessment—and readers will find examples of them within this book. Validity is not a dichotomous notion; instead, it spans a continuum. Validity is "the degree to which evidence and theory support the interpretations of test scores for proposed uses of the tests" (AERA et al., 2014, p. 11). Put simply, how confident can assessment users feel that the interpretations from an assessment's data can be trusted?

The *Standards* describe five sources of validity evidence: content, response processes, relations to other variables, internal structure, and consequences from testing. The amount of validity evidence needed to believe the outcomes and interpretations does not necessarily equate to evidence for all five sources, but it also does not equate to evidence from just one or two sources. For instance, decades of mathematics education research grounded validation arguments for assessments in test content (e.g., expert panel) evidence and/or internal structure (e.g., exploratory or confirmatory factor analysis). While a group of experts might agree that an assessment is connected to an intended construct, does it elicit the intended responses in appropriate ways (response process)? The factor structure and internal structure of an instrument might be satisfactory, yet the consequences from its use may have serious negative consequences that outweigh the benefits. Thus, assessment developers and users should consider some questions as they design and revise assessments:

1.   What is the intended use for the assessment?
2.   What evidence is needed to convey to others that the assessment does what it is designed to do?
3.   How should that evidence be gathered?

In that sense, some readers may sense connections between validation and backwards design (Wiggins & McTighe, 2005). If an assessment developer knows how the assessment will be used, for what purposes, and what interpretations may be drawn from assessment data, then the design decisions during assessment development are appropriately guided by a roadmap of sorts (AERA et al., 2014; Newcomer, 2012; Wilson, 2005; Wilson & Wilmot, 2019 [this volume]). Answering these questions may lead to gathering evidence for some but not all of the sources. Additionally, it is likely to lead to data collection during multiple rounds. For instance, assessment developers might gather evidence for content, response process, and internal structure during one round of data collection. A second or several subsequent, follow-up study or studies may take place to gather evidence related to relations to other variables and test consequences. These evidence pieces come together to form a validity argument. Walkowiak, Adams, and Berry (this volume) describe two separate studies within their chapter to provide evidence alongside two validation frameworks: Schilling and Hill's work (2007) and the *Standards* (AERA et al., 2014). Through multiple studies and a distinct focus on explicating the validity argument, Walkowiak and her colleagues provide one instantiation of how to combine frameworks to tell a comprehensive story about two assessments' outcomes. Moreover, their work suggests how assessment developers are not confined to using one validity framework. Instead, assessment developers might creatively weave evidence together across two frameworks to better ground their validity arguments regarding assessments' outcomes.

A validity argument is drawn together much like a mathematical proof, as a form of argumentative writing. These arguments are built on the complexity, quality, and frequency of validity evidence (AERA et al., 2014; Bostic, 2017; Kane, 2012, 2016; Jacobsen & Borowski, [this volume]; Lavery, Jong, Krupa, & Bostic, 2019; Walkowiak et al., 2019; Wilson & Wilmot, 2019 [this volume]). Wilson and Wilmot provide readers with a description of how the *Standards* (AERA et al., 2014) might be used in conjunction with the Bear Assessment System (Wilson, 2005) as a way to develop and frame validity arguments. There is neither one way to convey validity evidence to an intended audience nor one validation argument framework that is consistently better than others—it is up to the assessment developers to present their ideas in a coherent and consistent fashion (see Nilsson & Ryve, 2010 for a discussion of *coherence* and *consistency*). This volume contains several descriptions of how validity argument frameworks might be utilized to convey information about the results and conclusions from assessments.

We contrast argumentative writing with persuasive writing because one style of writing is more likely to be found within a validation argument. The former uses logical chains of ideas to communicate a rationale that claims and evidence are reasonable. Arguments may acknowledge limitations and share delimitations and are intended to convince a reader that ideas are sound. Persuasive writing, in contrast, intends to persuade a reader that one position is better than another in some regard. There is no other position to claim as being better when compared to an alternative; hence, persuasive writing is not purposeful for validation work. Argumentative writing aims to convince the reader using appropriate evidence and connections that an outcome was logically drawn. To that end, validity arguments offer readers a chance to thoughtfully consider an assessment developer's idea, its limitations, and discern the degree to which it is coherent. No validity argument is perfect; however, there are stronger ones based upon the frequency and type of evidence, and the quality of its purpose statement, given for an assessment.

Validation, as a process, provides developers and potential users with information about the degree to which one can trust the outcomes and interpretations of an assessment. It is essential for anyone engaging in quantitative research to consider the validity evidence and overall validation argument for an assessment that might be used. Kane (2016) reminds readers that validation should not be left to academics only—as those working in schools (e.g., teachers, curriculum specialists, principals, and other staff) should consider whether the information derived from an assessment logically flows from it. "Validation may not be easy, but it is generally possible to do a reasonably good job of [it] with a manageable level of effort" (Kane, 2016, p. 79). We agree that validation should be part of any study using an assessment meant to generate quantitative data, which includes but is not limited to surveys, concept inventories, measures, and observation protocols.

Validity evidence for validity arguments is tied to the outcomes and interpretations of an assessment; an assessment is not valid (AERA et al., 2014). An assessment is validated for use within particular contexts (i.e., settings, durations, and populations) and use outside of those specified contexts puts the interpretations at risk of being misinterpreted and inappropriately drawn (AERA et al., 2014; Kane, 2012; Newcomer, 2012). Validation is not something completed once and never examined again. It is a process, and validity arguments should be reevaluated when warranted (e.g., new population of respondents, revisions or updates to an assessment, or change in the level of stakes of interpretations). In this sense, as Jacobsen and Borowski (this volume) communicate, validation has merit as a methodology within mathematics education work. Similarly, Bostic, Matney, Sondergeld, and Stone (2019) argue that validation is akin to design-based research (e.g., Middleton, Gorard, Taylor, & Bannan-Ritland, 2008) because of their similarities. Generally speaking, validity arguments include a series of logical claims and inferences about the assessment. There are numerous frameworks for grounding validation frameworks (e.g., Kane, 2012; Mislevy, Almond, & Lukas, 2003; Pellegrino et al., 2016; Schilling & Hill, 2007; Wilson, 2005). This book provides examples of how assessment developers might overlay the *Standards* with published frameworks (see Walkowiak et al., this volume; Wilson & Wilmot, this volume). Because the sources of validity evidence function to some degree as buckets to fill (Lavery, Holloway-Libell, Amrein-Beardsley, Pivovarova, & Hahs-Vaughn, 2016), the *Standards'* expectation of gathering evidence to address sources of validity evidence meshes well with multiple validation argument frameworks. The key with any validity argument is that the necessary and sufficient evidence is provided. Weisstein (n.d.a) frames *necessary* as "a condition which must hold for a result to be true, but which does not guarantee it to be true." Similarly, Weisstein (n.d.b) characterizes *sufficient* as "a condition which, if true, guarantees that a result is also true." Thus, the decision about necessary and sufficient information to link claims and evidence within a validity argument for a particular assessment rests with the assessment developer's intentions. Hence, assessment development is not something to take lightly, perform quickly, or without purpose.

A central outcome from this book and its companion is filling a needed gap within the scope of mathematics education research with a resource for end-users and assessment developers describing quantitative measures for mathematics education. Assessment developers may feel more confident in their actions during the design phrases. Assessment users may feel empowered during the assessment selection process to think through their needs and assessments' intended uses—and to ask critical questions about the assessments they might use in practice and/or research. This book provides both theoretical narratives and practical examples about linking notions of validity with assessment construction and uses.

A second result from this project is to inform readers and draw together diverse scholars around shared interests. As evidenced from chapters in this volume and its companion, validation is best done with a group of scholars with a variety of expertise. This book sprung from ideas shared at a National Science Foundation–funded conference titled *Validity Evidence for Measurement in Mathematics Education* (V-M²Ed; award #1644314, #1644321). V-M²Ed brought together 40 individuals seeking to learn more and collaborate on issues related to validity evidence in mathematics education assessments and measures. Mathematics educators, educational psychologists, psychometricians, learning scientists, industry leaders, and graduate students from these fields joined together for two days of intense work in April 2017. A common sentiment shared by participants, as well as us, is that validation should not be left to a psychometrician or educational evaluator to perform reliability and factor analyses. Similarly, a content review conducted by a subject-matter expert functions as one piece of the validity argument for an assessment. Content review is often necessary but is not sufficient for robust interpretations from an assessment. Validation is a dynamic, flexible, and reasoned methodological approach, as Jacobsen and Borowski (this volume) suggest. Additionally, validation work is not purely quantitative work—in fact, much validation work is guided by qualitative research methods. Many chapters in this book, and its companion, discuss the plethora of qualitative research needed to support a robust validity argument for assessments and instruments. Measures of internal consistency and other psychometric modeling practices (e.g., item response theory, Rasch modeling, and structural equation modeling) are necessary but not sufficient for an assessment's outcomes and interpretations to have validity. These quantitative practices are ways to gather internal structure evidence and require a unique skill set to carry out. Thereby, validation may be conducted effectively by groups of individuals with differing skill sets, drawing together both quantitative and qualitatively minded scholars with the central aim of answering the question: How confident can assessment users feel that the outcomes from an assessment can be trusted? An outcome of such a partnership is that the qualitative vs. quantitative paradigm wars may be set aside for a greater, shared purpose.

A third applicable area for use of this book is within the communities of scholars and practitioners that are relatively underexplored in the mathematics education community. We intentionally included two chapters on statistics education topics as well as a call to action for mathematics coaching teacher educators to think about validation arguments. Related to statistics education, it has been well documented that statistics is an important component within K–20 educational contexts (Franklin et al., 2007; GAISE College Report, ASA Revision Committee, 2016). We purposefully chose two chapters among the submissions that focused on statistics education topics. Whitaker and colleagues (this volume) provide readers with an overview of a plan for a validity argument, which sets the stage for a research agenda over several years. Additionally, they take up the misuse of Cronbach's alpha has a measure of internal consistency and advocate for other statistical approaches. Readers wondering what a plan for conducting a rigorous validation study looks like should consult this chapter as one way it might occur. This chapter conveys what validation looks like when it is just beginning, which is a helpful contrast to Harrell-Williams' chapter that discusses results from years of work on an agenda focused on a validation argument for an assessment.

Harrell-Williams and her colleagues (this volume) explored instrument development that intends to capture secondary teachers' self-efficacy to teach statistics. Through this chapter, readers will learn about the instrument development process, decisions and choices made during it, and implications for its broad use. With the intention of showcasing assessment development at the early and late stages, we selected a piece by

Whitaker, Unfried, and Bond (this volume) to complement the other statistics education chapter. While Harrell-Williams' chapter describes a well-developed assessment and validity argument using a Rasch framework, Whitaker and colleagues characterize decision-making during the early stages of assessment design. Their chapter discusses choices and decision that must be made early on—definitions, construct delimitations and characteristics, frameworks, and intended uses of an assessment. It provides readers with important ideas about the potential challenges within assessment design and development. Moreover, it highlights an opportunity for scholars to engage in validation work without necessarily having to design and build an instrument from scratch: revisions and updates to previous instruments are the focus of their chapter. Such revisions and updates are necessary and part of the validation process when those modifications influence the validity argument and purpose statement (AERA et al., 2014). Revisions and updates should be taken up by mathematics (and statistics) education scholars and explored thoroughly through the validation process.

The second unique area, mathematics coaching, is a focused chapter by Harbour, Livers, and Hjalmarson (this volume) to advocate for greater need in reflecting on validity evidence related to assessments developed and used within mathematics coaching contexts. As Harbour and coauthors note, mathematics coaching scholarship is ripe for opportunities to develop and revise assessments that draw upon validation frameworks and have robust validity evidence. Many assessments and content measures have been used within coaching settings. Unfortunately, validity evidence and arguments may not necessarily be at the front of users' minds when selecting instruments for the purpose of mathematics coaching contexts, which includes mathematics coaches doing work within their districts and mathematics coaching teacher educators working with mathematics coaches. Therefore, conclusions from those assessments may be limited or not adequately drawn because of the limitations of the associated validity argument with the assessment. Put simply, this chapter provides a call to action for all who work in the mathematics coaching scholarly and practitioner areas.

In conclusion, we are confident that readers will find this volume, and its complement, valuable in supporting their future quantitative research with assessments and mathematics content measures. We intend for this introduction and accompanying chapters to spur new innovations in research into mathematics education phenomena. We aim to challenge the status quo within published, peer-reviewed mathematics education scholarship by being more inclusive when it comes validity evidence and validation arguments. A failure to include a discussion related to validity evidence for outcomes from an assessment within quantitative research is as bad as a failure to include a sample size of participants or description of data analysis procedures. Validity arguments for assessments' interpretations should be more prevalent within peer-reviewed mathematics education scholarship than they currently are. Taken collectively, we are both excited and optimistic for the future of assessment within mathematics education contexts.

References

American Educational Research Association, American Psychological Association, & National Council on Measurement in Education. (1999). *Standards for educational and psychological testing.* Washington, DC: American Educational Research Association.

American Educational Research Association, American Psychological Association, & National Council on Measurement in Education. (2014). *Standards for educational and psychological testing.* Washington, DC: American Educational Research Association.

Beckman, T. J., Cook, D. A., & Mandrekar, J. N. (2005). What is the validity evidence for assessments of clinical teaching? *Journal Of General Internal Medicine, 20*(12), 1159–1164.

Bostic, J. (2017). Moving forward: Instruments and opportunities for aligning current practices with testing standards. *Investigations in Mathematics Learning, 9*(3), 109–110.

Bostic, J., Krupa, E., & Shih, J. (2019). *Quantitative measures of mathematical knowledge: Researching instruments and perspectives.* New York, NY: Routledge.

Bostic, J., Krupa, E., Shih, J., & Carney, M. (2019). Reflecting on the past and thinking ahead in the measurement of students' outcomes. In J. Bostic, E. Krupa, & J. Shih (Eds.), *Quantitative measures of mathematical knowledge: Researching instruments and perspectives.* New York, NY: Routledge.

Bostic, J., Lesseig, K., Sherman, M., & Boston, M. (in press). Classroom observation and mathematics education research. *Journal of Mathematics Teacher Education.*

Bostic, J., Matney, G., Sondergeld, T., & Stone, G. (2019, February). Validation: A burgeoning methodology for mathematics education scholarship. In A. Sanogo & J. Cribbs (Eds.), *Proceedings of the 46th Annual Meeting of the Research Council on Mathematics Learning* (pp. 43–50), Charlotte, NC.

Boston, M., Bostic, J., Lesseig, K., & Sherman, M. (2015). Classroom observation tools to support the work of mathematics teacher educators. *Mathematics Teacher Educator, 3*, 154–175.

Franklin, C., Kader, G., Mewborn, D., Peck, R., Perry, M., & Schaeffer, R. (2007). *Guidelines for assessment and instruction in statistics education.* Alexandria, VA: American Statistical Association.

GAISE College Report ASA Revision Committee. (2016). *Guidelines for assessment and instruction in statistics education college report 2016.* Retrieved from www.amstat.org/education/gaise

Harbour, K. E., Livers, S. D, & Hjalmarson, M. A. (2019 [this volume]). Measurement and validity in the context of mathematics coaches. In J. Bostic,

E. Krupa, & J. Shih (Eds.), *Assessment in mathematics education contexts*. New York, NY: Routledge.

Harrell-Williams, L. M., Lovett, J. N., Lesser, L. M., Lee, H. S., Pierce, R. L., Murphy, T. J., & Sorto, M. A. (2019 [this volume]). Measuring self-efficacy to teach statistics in grades 6–12 mathematics teachers. In J. Bostic, E. Krupa, & J. Shih (Eds.), *Assessment in mathematics education contexts: Theoretical frameworks and new directions*. New York, NY: Routledge.

Hill, H., & Shih, J. (2009). Examining the quality of statistical mathematics education research. *Journal of Research in Mathematics Education, 40*(3), 241–250.

Jacobsen, E., & Borowski, R. (2019 [this volume]). Measure validation as a research methodology for mathematics education. In J. Bostic, E. Krupa, & J. Shih (Eds.), *Assessment in mathematics education contexts: Theoretical frameworks and new directions*. New York, NY: Routledge.

Kane, M. T. (2001). Current concerns in validity theory. *Journal of Educational Measurement, 38*(4), 319–342. doi:10.1111/j.1745-3984.2001.tb01130.x

Kane, M. T. (2012). All validity is construct validity: Or is it? *Measurement: Interdisciplinary Research and Perspectives, 10*(1–2), 66–70.

Kane, M. T. (2016). Validation strategies: Delineating and validating proposed interpretations and uses of test scores. In S. Lane, M. R. Raymond, T. M. Haladyna, S. Lane, M. R. Raymond, & T. M. Haladyna (Eds.), *Handbook of test development* (2nd ed., pp. 64–80). New York, NY, US: Routledge & Taylor & Francis Group.

Krupa, E., Carney, M., & Bostic, J. (2019). Approaches to instrument validation. *Applied Measurement in Education, 32*(1), 1–9.

Lavery, M. R., Holloway-Libell, J., Amrein-Beardsley, A., Pivovarova, M., & Hahs-Vaughn, D. (2016). *Evaluating the validity evidence surrounding the use of student standardized test scores to evaluate teachers: A centennial, systematic mega-review.* Paper presented at the American Educational Research Association Annual Meeting, Washington, DC. Retrieved from www.aera.net/Publications/Online-Paper-Repository/AERA-Online-Paper-Repository

Lavery, M. R., Jong, C., Krupa, E., & Bostic, J. (2019 [this volume]). Developing an instrument with validity in mind. In J. Bostic, E. Krupa, & J. Shih (Eds.), *Assessment in mathematics education contexts: Theoretical frameworks and new directions*. New York, NY: Routledge.

Lissitz, R. W., & Samuelsen, K. (2007). A suggested change in terminology and emphasis regarding validity and education. *Educational Researcher, 36*(8), 437–448.

Middleton, J., Gorard, S., Taylor, C., & Bannan-Ritland, B. (2008). The "compleat" design experiment. In A. Kelly, R. Lesh, & J. Baek (Eds.), *Handbook of design research methods in education: Innovations in science, technology, engineering, and mathematics teaching and learning* (pp. 21–46). New York, NY: Routledge.

Mislevy, R. J. (2007). Validity by design. *Educational Researcher, 36*(8), 463–469.

Mislevy, R. J., Almond, R. G., & Lukas, J. F. (2003). *A brief introduction to evidence-centered design*. Retrieved from www.ets.org/Media/Research/pdf/RR-03-16.pdf

Newcomer, K. (2012). Basics of design for evaluation of cohesion policy interventions. In K. Olejniczak, M. Kozak, & B. Bienias (Eds.), *Evaluating the effects of regional interventions: A look beyond current structural funds practice* (pp. 161–176). Republic of Poland: Ministry of Regional Development.

Nilsson, P., & Ryve, A. (2010). Focal event, contextualization, and effective communication in the classroom. *Educational Studies in Mathematics, 74*(3) 241–258.

Pellegrino, J., DiBello, L., & Goldman, S. (2016). A framework for conceptualizing and evaluating the validity of instructionally relevant assessments. *Educational Psychologist, 51*(1), 59–81.

Schilling, S., & Hill, H. (2007). Assessing measures of mathematical knowledge for teaching: A validity argument approach. *Measurement: Interdisciplinary Research and Perspectives, 5*(2–3), 70–80.

Walkowiak, T. A., Berry, E. A., & Berry, R. Q. (2019 [this volume]). Validity arguments for measures of mathematics teaching practices: Comparing the M-SCAN and IPL-M. In J. Bostic, E. Krupa, & J. Shih (Eds.), *Assessment in mathematics education contexts: Theoretical frameworks and new directions.* New York, NY: Routledge.

Weisstein, E. (n.d.a). *Necessary*. Retrieved from http://mathworld.wolfram.com/Necessary.html

Weisstein, E. (n.d.b). *Sufficient*. Retrieved from http://mathworld.wolfram.com/Sufficient.html

Whitaker, D., Unfried, A., & Bond, M. (2019 [this volume]). Design and validity arguments for the surveys of motivational attitudes toward statistics (SOMAS) instruments. In J. Bostic, E. Krupa, & J. Shih (Eds.), *Assessment in mathematics education contexts: Theoretical frameworks and new directions.* New York, NY: Routledge.

Wiggins, G., & McTighe, J. (2005). *Understanding by design*. Alexandria, VA: Association for Supervision and Curriculum Development.

Wilson, M. (2005). *Constructing measures: An item response modeling approach.* Mahwah, NJ: Erlbaum.

Wilson, M., & Wilmot, D. B. (2019 [this volume]). Gathering validity evidence using the BEAR assessment system (BAS): A mathematics assessment perspective. In J. Bostic, E. Krupa, & J. Shih (Eds.), *Assessment in mathematics education contexts: Theoretical frameworks and new directions.* New York, NY: Routledge.

Ziebarth, S., Fonger, N., & Kratky, J. (2014). Instruments for studying the enacted mathematics curriculum. In D. Thompson & Z. Usiskin (Eds.), *Enacted mathematics curriculum: A conceptual framework and needs* (pp. 97–120). Charlotte, NC: Information Age Publishing.

2 Developing an Instrument With Validity in Mind

Matthew R. Lavery, Cindy Jong,
Erin E. Krupa, and Jonathan D. Bostic

Educational and psychological testing and assessment are among the most important contributions of cognitive and behavioral sciences to our society. Not all tests are well-developed, nor are all testing practices wise or beneficial, but there is extensive evidence documenting the usefulness of well-constructed, well-interpreted tests. Well-constructed tests that are valid for their intended purposes have the potential to provide substantial benefits for test takers and test users.

(American Educational Research Association [AERA], American Psychological Association [APA], & National Council for Measurement in Education [NCME], 2014, p. 1)

These opening words of the most recent edition of the jointly published *Standards for Educational and Psychological Testing* (AERA et al., 2014; henceforth referred to as the *Standards*) suggest the centrality and importance of measurement to society. Not only is sound measurement important for the conduct of cognitive and behavioral science, but the instruments themselves and improvements in measurement developed within these scientific fields have made valuable contributions. Educational and psychological testing uses instruments designed to quantify educational and psychological constructs. In this chapter, we define a construct as it is defined in the *Standards*, as "the concept or characteristic that a test is designed to measure" (AERA et al., 2014, p. 11). The *Standards* give "mathematics achievement, general cognitive ability, racial identity attitudes, depression, and self-esteem" (p. 11) as examples of currently assessed constructs. There are numerous constructs that may fall into the cognitive, affective, or psychomotor domains (see Anderson & Krathwohl, 2001; Bloom, 1956; Krathwohl, 2002). Similarly, the instruments developed to measure them are equally varied and may include knowledge assessments that look like traditional tests (whether delivered on paper or via computer), surveys or questionnaires designed to measure affective constructs, or observation protocols that employ checklists or rubrics to code or quantify observed evidence of a construct. These instruments can vary in precision and quality, and it is incumbent upon developers to provide evidence that instruments perform adequately for their intended use (AERA et al., 2014).

The reader should note two subtleties of the prior assertion. First, while instruments may vary in precision and quality, it is the interpretation or use of the instrument that must be validated, rather than the instrument itself (AERA et al., 2014). Second, the definition of "adequate" may vary based on the specific interpretation or use being validated. Kane (2013) writes that "more-ambitious interpretations and uses require more backing (i.e., evidence) than less-ambitious interpretations and uses" (p. 21; see also AERA et al., 2014, p. 13). If, for example, a three-minute, online quiz assigns someone to the wrong fictional wizarding house, the consequences of that error are far less grave than if an exam grants licensure to perform life-saving medical procedures to wholly unqualified test takers. Trivial tests may not require very rigorous validation, but assessments that inform high-stakes decisions, or decisions for which the consequences of "getting it wrong" are either grave or costly, require that substantially more time, attention, and resources be devoted to ensuring they produce accurate scores that are valid for the intended interpretations and uses.

While it is true that validity theory and validation have a long and nuanced history (see Haertel & Herman, 2005), it may not be necessary for all instrument development projects to include a validation expert who has studied this literature extensively. Many scholars can construct an instrument for use within their own field that performs sufficiently well for its intended purpose and collect enough evidence to document its validity for that purpose by carefully attending to certain key questions during its development. We do not suggest that instrument development and validation is a simple exercise. We echo the sentiment of Kane (2016), who writes that "validation may not be easy, but it is generally possible to do a reasonably good job of [it] with a manageable level of effort" (p. 79). We also do not suggest that special expertise or specific skills are never required for this work (quite the contrary, in fact, as we discuss in a forthcoming section on assembling the development team). The premise of this chapter is that instrument developers should make careful, conscious decisions throughout the planning and development process to produce a robust validity argument supported by sufficient evidence for its proposed interpretations and uses. This premise is consistent with the idea that Briggs (2004) refers to as design validity or that Mislevy and Riconscente (2006) discuss as a central aspect of evidence-centered design.

The purpose of this chapter is to present some key questions that scholars should consider during three phases of instrument development (i.e., planning, production, and piloting), describe some of the design decisions implied by possible answers to those questions, and discuss some of the analyses and procedures that scholars can use to test the performance of the instrument and collect evidence that it is valid for its intended use. We note that there are many approaches to and frameworks for validation that have been described in the literature (see Kane, 2004; Mislevy & Riconscente, 2006; Pellegrino, DiBello, & Goldman, 2016; Schilling, 2004; Wilson, 2005; see also Wilson & Wilmot, this volume), and this chapter is not meant to favor any one of them over the others. However, we recognize the *Standards* (AERA et al., 2014) as an authoritative reference on this topic, and will connect our recommendations to the five sources of validity evidence (evidence based on test content, evidence based on response processes, evidence based on internal structure, evidence based on relations to other variables, and evidence based on the consequences of testing; pp. 13–21), and the foundational concerns of reliability/precision and fairness, described therein. Each section of this chapter will discuss three examples to focus the reader's attention on the differences in the evidence required to build validity arguments for their different purposes. Each forthcoming section explores the key questions and related validity evidence from three different measurement perspectives: assessment, evaluation, and research. We do not select these examples as endorsements of any particular instruments, nor do we intend these examples to be exhaustive or comprehensive. Our examples have simply been selected as a starting point to support discussion of several common concerns. Readers seeking research ideas should explore chapters in this book and its companion, *Quantitative measures of mathematical knowledge: Researching instruments and perspectives* (Bostic, Krupa, & Shih, 2019).

Assessing Knowledge of Equivalence

We first consider the mathematics classroom where instruments are regularly developed, administered, interpreted, and applied. The inferences and decisions supported by classroom assessments are different than many of the other applications of psychological measurement, but that does not mean that they are any less complex. Classroom teachers, and those who develop instructional assessments for use in classroom settings, must still consider validity concerns, but should often do so with greater efficiency, since these assessments may be developed with limited time and resources, and without large samples of participants to complete pilot versions of the test (Gerber, Bostic, & Lavery, 2018; Gerber, Lavery, & Bostic, in press). In our classroom assessment example, we explore a formative assessment used to inform instructional decisions about student knowledge of equivalence. We acknowledge that classroom assessments are also commonly used to determine student grades but will not discuss grading practices in the present chapter. The interested reader is encouraged to explore the works of Brookhart and Guskey on this topic, especially Brookhart et al. (2016) for a review of research on grading. In this chapter, we will describe the hypothetical development of a formative assessment of students' understanding of equivalence. We intentionally use a hypothetical formative assessment for our example so that we can discuss some of these development concerns from the perspective of a classroom teacher.

Evaluating Professional Development on the Standards of Mathematical Practice (SMPs)

Researchers may also develop instruments to evaluate initiatives and programs that are designed to increase teachers' knowledge and skills and improve instruction in order to realize better outcomes for students. Generally, these programs include multiple teachers and their students, and the instruments developed for program evaluation and improvement purposes might not be meant to contribute to theory or to generalize beyond the participants assessed. In this chapter, we will discuss an existing instrument that can be used for program evaluation purposes and which an author of this chapter has developed and validated, the Revised SMPs Look-for Protocol (henceforth, Revised SMP Protocol; Bostic, Matney, & Sondergeld, 2019; Bostic & Matney, 2016). In this case, we use the actual example of the Revised SMPs Protocol to allow discussion of some unique concerns related to observation protocols based on the author's experience.

Researching Self-Efficacy

Finally, mathematics education researchers may also develop instruments to measure a variety of constructs (e.g., attitudes and beliefs about mathematics, teaching practices, students' mathematical thinking). Research instruments can be designed to inform instruction or to evaluate programs and interventions, but here we use an example designed to help researchers understand a construct, build theory, or generalize knowledge beyond the participants or context measured. For our research example, we will discuss the Mathematics Teaching Efficacy Beliefs Instrument (MTEBI; Enochs, Smith, & Huinker, 2000). Although we were not involved in the development of the MTEBI or validation of its use, we have found sufficient illustrative examples in the papers written by its developers for the purposes of this chapter.

Start at the Beginning With the End in Mind

The first, and possibly most important, phase of instrument development is the planning phase. Though we could choose any one of a number of metaphors to illustrate this point (such as the dangers of constructing a building without a blueprint), we prefer to leave the importance of planning to the common sense of the reader. This allows us to instead focus our attention on the single most critical validation concern for all instruments, which lies at the center of the planning phase; that it is the intended interpretation and use of the instrument that is validated, and not the instrument itself (AERA et al., 2014; Kane, 2013; Messick, 1989). Instruments are not inherently well-designed so much as they are well-designed for a particular purpose. Defining that purpose as thoroughly and clearly as possible first will focus and guide the efforts of developers through all of the steps that follow. Developers should draft a concise statement that summarizes the intent of the instrument before any of its components or items are crafted.

The summary statement described in this chapter, including the seven questions discussed next, is based on the recommendations offered by the 40 participants at Validity Evidence for Measurement in Mathematics Education (V-M²Ed), an NSF-sponsored conference of experts in mathematics education, psychometrics, and validation (V-M²Ed participants, personal communication, April 2–3, 2017; see http://measuresinmathed. org/v-m2ed-conference/). A clearly articulated summary statement will ultimately help other mathematics educators use the instrument in ways that are supported with validity evidence, and not apply it to purposes for which it has not been designed or tested. By crafting the summary statement at the beginning of the process, developers can ensure that they think about the intended uses for which they plan to design the instrument (and must therefore collect evidence to support its validity for that purpose), and which uses of the instrument are outside its scope (and therefore need not be validated). The summary statement should answer the following questions as clearly and succinctly as possible:

1. What construct is measured by the instrument?
2. How is the construct being measured?
3. Who is being measured?
4. In what context, setting, and under what conditions is the instrument used?
5. How and by whom are scores calculated and interpreted?
6. How should the instrument scores be interpreted and used?
7. What interpretations and uses are inappropriate for the scores?

The above questions may seem deceptively simple at first read, but each has extensive implications for the development and validation of a measurement instrument. Once such a summary statement is drafted in detail, it guides the development of the validity argument in which developers collect the evidence necessary to convince the reasonable critic that the instrument developed is valid for the purpose and use described.

Thus, as questions 1–7 are meant to define the intended purpose of the instrument, each of them is immediately followed with three secondary questions: (1) "What evidence must I collect to convince a reasonable critic that this instrument adequately fulfills the purpose implied by this question?" (2) "What challenges might a reasonable critic raise related to this question (such as possible sources of error or bias)?" and (3) "What evidence must I collect to show that the instrument is robust to those challenges?" Note that until an instrument is put into active use, it may often be more productive to brainstorm possible ways that an instrument might fail and to collect evidence that it has not done so, than it is to try to show that it accomplishes its goals. To accumulate positive evidence that an instrument works as intended, developers may need to collect much more evidence, using multiple group designs, than is typically feasible during the development process (for a concise discussion of this paradox with certification tests, see the opening paragraphs of Briggs, 2004).

These questions are discussed in more detail in terms of the construct, context, and purpose of the instrument, but consideration of each of these questions is often informed and influenced by answers to the other questions. Throughout this chapter, we may discuss instrument development in a seemingly linear and sequential fashion, but this is done to illustrate how these considerations build and depend upon one another. In practice, however, developers may find this process to be more dynamic, iterative, and cyclical, as questions explored, work done, and data collected at each step often have implications for other steps, even if they were already addressed. This is another reason to plan thoroughly, as good planning can often help prevent developers from learning something later in the process that requires them to duplicate or discard earlier efforts.

Map Out the Construct, Content, or Skill

An early step in developing an assessment that performs its specified purpose well with the specified population is to clearly conceptualize the construct, content, or skill being measured. Questions 1 and 2 of the summary statement can be more fully explored through the development questions shown in Table 2.1. Answering these questions typically begins with a thorough review of the scholarly literature surrounding that which is being measured. When measuring content knowledge, skill mastery, or any construct for which it makes sense for respondents to have "more" or "less" of it (such as self-efficacy or motivation), it is important to not only conceptualize that which is measured at all relevant points on its continuum. Sometimes, it is important to measure the variance among test takers, either to compare or classify them, or to understand their progress in its development. In such cases, developers must conceptualize the construct, content, or skill along the full continuum that they intend to measure. If there is a target level (a level at which test takers

Table 2.1 Development Questions Regarding the Construct to Consider During the Planning Phase

Questions	Implications for Development/Validity Evidence
1.a Is the construct fixed, will it change steadily over time (i.e., consistently increase, decrease, or develop in predictable way), or will it fluctuate?	• Scores on repeated administrations for the same participants should follow expectations and reflect theory (validity evidence based on test content, internal structure, and reliability/internal consistency).
1.b What might responses look like along a continuum of the construct, or at various levels of understanding?	• If the instrument should distinguish between respondents who differ in status, then it must have items that reflect each status of interest (validity evidence based on test content, internal structure, and reliability/internal consistency).
1.c How are respondents expected to use or apply the content, or to manifest or demonstrate construct?	• When producing items, consider whether and how the instrument should resemble real-life application or use (validity evidence related to test content and response processes).
1.d What responses can be expected from respondents below the target level of understanding, or at a status other than the status of interest?	• When producing items, consider developing items to target incomplete understandings or misconceptions, or to differentiate among construct statuses (validity evidence related to test content and response processes).
1.e What constructs, knowledge, or skills are related to the measured construct (perhaps as prerequisites or corequisites)?	• When producing items, consider crafting them to measure prerequisite/corequisite skills/constructs, or to avoid measuring them, depending on the proposed interpretation and use (validity evidence related to test content and response processes).
1.f How will the content or construct be further used, developed, or built upon in the future?	• When producing items, consider whether and how to incorporate related or advanced content or to foreshadow future uses, based on the proposed interpretation and use (validity evidence related to test content and response processes).
1.g What misconceptions or ideas are common for those who have not yet reached the desired target level? What behaviors, views, or other manifestations are common for those who have not yet fully attained the status of interest with the construct?	• When producing items, consider developing items to target incomplete understandings, misconceptions, or early stages of development (validity evidence related to test content and response processes).
2.a How is the construct most directly and most accurately observed?	• When producing items, consider designing them to resemble authentic or direct performances (validity evidence related to response processes).

2.b Should the instrument measure only end results or final answers, or is it important to observe the process that leads to them, as well?	• If it is important to measure the process, then the instrument must have items designed to capture the process (validity evidence related to test content and response processes).

Note: Development questions 1.a through 1.g relate to summary statement question 1, "What construct is measured by the instrument?" and development questions 2.a and 2.b relate to summary statement question 2, "How is the construct being measured?"

are expected to exhibit or possess the measured construct, content, or skill), developers must fully define that level, as well as the stages before and after it. It is also critical to assess prerequisites for achieving the target level, and common misunderstandings or misapplications of those approaching the target level, in order to best understand how to measure respondents at various levels of the construct. During this step, it may help developers to locate or draft a *construct map*, defined as "a well-thought-out and researched ordering of qualitatively different levels of performance" (Black, Wilson, & Shih-Ying, 2011, p. 94; see also Maloney, Confrey, & Nguyen, 2014; Wilson, 2005). Although these references discuss construct maps in terms of knowledge assessments, similar orderings of qualitatively different levels of any construct could be constructed to inform affective measures or observation protocols, as well as conceptual models or theoretical frameworks (discussed in the Researching Self-Efficacy section).

Assessing Knowledge of Equivalence

If a third-grade mathematics teacher is planning to teach a lesson on solving for an unknown number as part of a unit on algebraic thinking, the teacher might be interested in assessing students' understanding of the equal sign and the concept of equality to build on what the students already know and uncover any misconceptions. Carpenter, Franke, and Levi's (2003) book on integrating arithmetic and algebra includes formative assessments and insights into common ideas students have about the equal sign that can serve as a guide in planning. For example, students can complete an open number sentence such as $8 + 4 = \_\_ + 5$ and describe what the equal sign means. This brief classroom assessment, which could take multiple forms depending on the students and grade level (e.g., formative paper-and-pencil task, whole-class discussion, diagnostic one-on-one interview), could then be used to determine which students understand equality and can explain that the equal sign acts as a balance or a relation between two numbers. The teacher can also determine that some students have a common narrow view of the equal sign as a symbol for calculation that is followed with an answer.

In formative assessments like these, teachers are typically interested not only in whether students answer the questions they ask correctly, but also observing how students think about the questions and observe the process that they use to arrive at an answer. A well-developed construct map could help the teacher identify and look for specific evidence of early, incomplete, or fragile understandings, and determine the logical next instructional step that is most likely to move students toward mastery of the concept.

Evaluating the SMPs

Evaluators wished to explore the effectiveness of a grant-funded professional development program designed to support K-12 teachers' knowledge and practices related to the SMPs. The SMPs are mathematical behaviors and habits that are associated with the National Council of Teachers of Mathematics (NCTM) process standards (NCTM, 2000) and the ideals of mathematical proficiency (Kilpatrick, Swafford, & Findell, 2001). Teachers are expected to promote the SMPs during instruction with the aim that students are more likely to engage in them (Bostic & Matney, 2013). During a yearlong professional development program focused on the SMPs and K–5 mathematics content, the evaluation included two measures: the Standards of Mathematical Practice Knowledge Assessment (SMP-KA; Matney, Bostic, & Lavery, 2019) and the Revised SMPs Look-for Protocol. The Revised SMPs Protocol can be used to measure the degree to which teachers promote the SMPs during classroom instruction (Bostic et al., 2019; Bostic & Matney, 2016). Development of the professional development program required creation of a construct map describing qualitatively different levels of student engagement with the SMPs, as well as a construct map describing levels of teachers' knowledge and skills promoting the SMPs. These parallel construct maps were used to create a logic model that effectively assessed what teachers knew before and after the professional development program, and the degree to which they promoted the SMPs during classroom instruction before and after the program, informing the creation of both instruments.

Researching Self-Efficacy

The measurement needs for research are often very different from the purposes for which classroom assessment and evaluation instruments are designed. Research questions tend to rely on theory more than the other two measurement perspectives. Thus, researchers may find conceptual models or theoretical frameworks to be more useful organizational tools than construct maps. For example, when investigating teachers' self-efficacy, the researcher may also wish to measure vicarious experiences, social persuasion, and emotional arousal, the other major sources of efficacy information identified by Bandura (1977; see the conceptual model provided in Figure 2, p. 195). Using the framework from Bandura and developed by modifying the Science Teaching Efficacy Beliefs Instrument (STEBI-A; Riggs & Enochs, 1989), Enochs et al. (2000) created the MTEBI to provide a quantitative measure of teaching efficacy beliefs to provide a picture of mathematics teaching behavior.

Additional Concerns

When mapping the construct to be measured, developers should consider how well-developed the theory surrounding it is. For example, the theoretical framework around sense of self-efficacy is well-developed and built on a long and rich history of scholarly literature, allowing researchers interested in this construct to build a thorough understanding of it through an extensive review of the literature. When theory surrounding the construct of interest has not yet been fully developed, researchers might be better served in further theory-building rather than instrument development. Alonzo (2007) cautions that prematurely building an instrument to measure a construct runs the risk of codifying, and thus "locking in," an underdeveloped theoretical framework. Conversely, some argue that instrument development can serve an important role in developing our understanding of a construct (see Carney, Totorica, Cavey, & Lowenthal, 2019). In either case, it is most important that developers investigate and understand the condition of current theory surrounding the construct measured, and thoughtfully consider their role in advancing theory through development of the instrument.

Consider the Target Population, Context, and Setting

Summary statement questions 3–5 speak to the importance of context in the instrument development process and can be explored more fully through the development questions shown in Table 2.2. Instruments developed and tested for use with one population in a particular context do not necessarily support valid inferences and decisions with a different population, or even with the same population in a different context. A test of content knowledge designed to be administered to teachers may not be suitable for use with children. Similarly, if that instrument was designed to be a timed test administered in controlled conditions with no reference material, the scores produced if teachers took the test online at their own pace could not be interpreted the same way. As such, developers must consider for which test takers and in which contexts they will develop the instrument so that they can collect sufficient evidence that it works as designed. Further, if an instrument is used in a different context or population than intended, additional evidence needs to be gathered,

Table 2.2 Development Questions Regarding the Context to Consider During the Planning Phase

Questions	Implications for Development/Validity Evidence
3.a To whom will the instrument be administered?	• When piloting, the pilot sample must be representative of the intended population for validity evidence to generalize.
3.b Is the instrument being used to measure something about the test takers themselves, or about some other level or unit (e.g., a teacher or school)?	• Consider what information informs the instrument's proposed interpretation/use and from whom to collect it (validity evidence related to response processes).
3.c What resources and supports are necessary for members of the target population in this context, and what resources and supports should be disallowed?	• Has implications for standardizing administration procedures, and determining appropriate accommodations (implications for reliability/internal consistency).
3.d Are there any possible confounding factors associated with the population of interest, or that could be associated with some test takers?	• Consider how to test for potential sources of bias and how to mitigate or correct for them if found (validity evidence based on internal structure).
3.e Are there any test takers for whom the instrument being developed would be inappropriate?	• Validity evidence must be collected for the instrument's proposed interpretations and uses; summary statement should explicitly exclude populations not vetted.
4.a Is the instrument being developed for only one instance of a context (e.g., a teacher's geometry class), for multiple instances of similar contexts (e.g., all geometry classes in a school or district), or for multiple contexts (e.g., any student learning geometry anywhere)?	• When producing items, must consider the possible influence of the various backgrounds, prior knowledge, and contexts that test takers bring to the instrument (validity evidence related to response processes).
4.b Are there any contexts for which the instrument being developed would be inappropriate?	• Validity evidence must be collected for the instrument's proposed interpretations and uses; summary statement should explicitly exclude contexts not vetted.
4.c How will the instrument be administered, and responses collected?	• Has implications for standardizing administration procedures, and determining appropriate accommodations (implications for evidence of reliability/internal consistency).
5.a Is special training required to administer, score, and/or interpret the instrument?	• Has implications for standardizing administration and scoring procedures (implications for evidence of reliability/internal consistency).

| 5.b Are the decisions informed by the instrument made by those who will score and interpret the assessment? | • Has implications for standardizing administration and scoring procedures; may need to create detailed interpretation guidelines. |

Note: Development questions 3.a through 3.e relate to summary statement question 3, "Who is being measured?"; development questions 4.a through 4.c relate to summary statement question 4, "In what context, setting, and under what conditions is the instrument used?"; and development questions 5.a and 5.b relate to summary statement question 5, "How and by whom are scores calculated and interpreted?"

or potentially a new validation study altogether. When answering these development questions, it is often helpful for developers to think through the administration of the instrument as intended, and then to consider any possible variations of that hypothetical administration to determine how these variations might affect the design process.

Assessing Knowledge of Equivalence

The teacher described prior, who aims to assess student understandings of the equal sign and equality, is certainly interested in administering the assessment to their own students and possibly collaborating with other teachers at the school, but due to time constraints would likely want to select items that are capable of producing responses that represent ideas common among elementary-aged students. For this intended use, it is important to carefully construct the assessment, and to collect evidence (e.g., student work samples) to appropriately inform instructional decisions. Consider if this classroom teacher used a poorly constructed assessment item such as "5 + 5 = __ + 0." The teacher would not be able to discriminate whether students understood the concept of equality or what the equal sign meant, because students would likely answer 10 without thinking about the zero. It could provide misleading information and lead the teacher to make invalid inferences about student understanding and instructional needs. Such incorrect inferences might lead the teacher to plan inappropriate instruction, based on an item that was too easy or limited in its ability to show variability in students' thinking. Thus, even formative assessments delivered at the classroom level can benefit from considering validity of the intended interpretations.

Evaluating the SMPs

The Revised SMPs Protocol (Bostic et al., 2019; Bostic & Matney, 2016) is designed to gather evidence related to teachers' promotion of the SMPs and may be used with live or video-recorded classroom instruction. Teachers could potentially demonstrate their understanding of how to promote the SMPs during instruction by adding statements related to the

SMPs to their lesson plans, or by teaching a simulated lesson with other teachers during a professional development session in order to demonstrate what they might do with their students. However, both of these instances may be too abstract to inform a valid evaluation of the professional development program and would not necessarily indicate the degree to which a teacher has changed instructional practices. Thus, the Revised SMPs Protocol is used with observed instruction because it provides a clearer picture of teachers' actual practice in classroom settings.

Researching Self-Efficacy

Instruments developed for research purposes are often intended to produce generalizable results. Thus, while individual scores are produced for individual participants, those participants may be selected to act as a representative sample of a particular target population. This is an especially relevant concern when the respondents available during development may be systematically different or may fail to adequately represent, the target population. For instance, the MTEBI (Enochs et al., 2000), was a subject-specific instrument used to measure teacher beliefs. During the validation study of the MTEBI the researchers had access to preservice elementary mathematics teachers, and thus were only able to collect validity evidence to support its use with a similar target population. Researchers need to be aware that the participants in their validation study might not adequately represent their intended target population of all mathematics teachers however, as their knowledge about and experience with teaching mathematics may be very limited. It is reasonable to expect that practicing teachers, who have more extensive and more recent experiences relevant to the study, may respond differently than preservice teachers do. Thus, if the target population for a research instrument is practicing teachers, then practicing teachers need to be tested.

Additional Concerns

Across all three contexts developers must also think about special cases of test takers and how the design of the instrument may affect its fairness to them (see *fairness* in AERA et al., 2014). For example, if an instrument being described here were to use idiomatic language or culturally situated examples, the phrasing of these items might introduce measurement error when taken by participants for whom English is a second language or who may have been raised in a different culture and then migrated to the area where the instrument is used. Similarly, some test takers may possess disabilities, unrelated to the construct measured, that interfere with their ability to understand and respond to the instrument as intended. Though insufficient reading level may be less of a concern for researchers studying practicing teachers, students who struggle in reading may not be able to demonstrate their true levels of content mastery in mathematics on a written assessment if the questions rely too heavily on text written above students' reading levels. Concerns such as these can and should affect both the design and validation of the instrument in question.

Define the Intended Purpose

As mentioned prior, instruments are not validated, their interpretations and uses are (AERA et al., 2014; Kane, 2013; Messick, 1989). As such, a critical step in planning development of an instrument is to thoroughly consider summary statement questions 6 and 7. By carefully defining the intent of the instrument, developers are able to plan, in advance, what kinds of validity evidence they need to collect and how to collect it. When considering the purpose of the instrument, developers should consider development questions such as those shown in Table 2.3. These questions may seem relatively straightforward, but they have serious implications for instrument development and validation, and may have serious implications for test takers, as well. If an instrument is being designed for use as a certification exam or an admissions test, then the consequences to the test taker are directly based on the score that the instrument produces and increase the need for evidence that it produces accurate, unbiased scores that are fair when used for that purpose.

Assessing Knowledge of Equivalence

In the case of formative classroom assessment, the purpose is for teachers to understand students' needs so that they can provide the most appropriate instruction. The instrument informs instructional decisions. As such, the purpose is not to rank the students by concept mastery or to compare each student to some predetermined cutoff score. Instead, the information needs to be instructionally relevant; ideally it is both detailed and diagnostic. In our example, the assessment should help the teacher understand where the students are in the development of their understandings of the equal sign and equality. If the assessment only gives the teacher a total number of correct responses, without providing some substantive information about student understanding, it does not serve an instructionally relevant purpose. Since students explain their thinking as part of the assessment, the teacher can expect to detect if a student gives a correct answer based on incorrect reasoning, or the reverse.

Evaluating the SMPs

The Revised SMPs Protocol is also designed to support formative decisions, when used with individual teachers, or can be used to evaluate a professional development program when aggregated across participants

Table 2.3 Development Questions Regarding the Intended Interpretation and Use to Consider During the Planning Phase

Questions	Implications for Development/Validity Evidence
6.a What types of decisions or actions will be based on the information (i.e., scores) that the instrument provides?	• Validity argument must establish that the instrument's scores provide the information necessary to make valid decisions and take appropriate actions (implications for presentation of the validity argument).
6.b Will the instrument be used to establish a minimum level of competency, or to determine competency on a continuum?	• Scores must be sufficiently precise at the critical decision point for the decisions in question to be consistently valid (i.e., a test of minimum competency need not accurately measure degrees of advanced competency; evidence of reliability/precision and internal consistency).
6.c Will the instrument classify, sort, rank, or compare test takers to one another?	• Instrument must be accurate at all levels of competency and measurement error must be minimized (so that confidence intervals are smaller) to rank, sort, or compare (evidence of reliability/precision and internal consistency).
6.d Will the instrument support diagnostic analyses by measuring components of the construct, or levels of subconstructs, separately?	• Components and subconstructs of interest must be measured independently of one another for meaningful diagnostic analyses (validity evidence based on internal structure).
6.e Will there be positive or negative consequences that will be associated with use of the instrument (e.g., will scores determine access to resources or programs)?	• During implementation of the completed instrument, investigate that intended consequences are realized without incurring negative unintended consequences (validity evidence based on related consequences).
6.f Will the instrument be scored against externally defined criteria or existing standards?	• Scores from the developed instrument should correlate with other existing measures the same standards or criteria, if applicable (validity evidence based on relations to other variables). • If an external authority has set the relevant criteria or standards, officials from that authority or experts on their standards can review the instrument for appropriate alignment (validity evidence based on test content).

Note: Development questions 6.a through 6.f relate to summary statement question 6, "How should the instrument scores be interpreted and used?" Summary statement question 7, "What interpretations and uses are inappropriate for the scores?" must be considered throughout the development and validation process and has been reflected in the development questions presented in Tables 2.1–2.3.

(Bostic et al., 2019). The protocol is a unidimensional observation protocol with several indicators for each of the eight SMPs. Pulling one indicator and searching for that one aspect would not support diagnostic evaluations. Moreover, results only indicate whether evidence was observed; it does not measure degree of the construct and cannot be parsed into a short form. Decisions related to results from the protocol should be connected to its use in instructional contexts. The tool is not meant for high-stakes evaluation; instead, its outcomes should either generate conversations with teachers or be used to demonstrate changes in instruction over a period of time. Since decisions informed by results from the protocol may ultimately be made by others, it is centrally important that instrument users provide adequate discussion of the results to others, such as school district personnel and educational stakeholders beyond the classroom teacher and qualify that decisions should not be punitive or lead to preferential treatment.

Researching Self-Efficacy

The MTEBI instrument is by its nature diagnostic, in that the researcher's intent is to parse out one source of influence on participants' senses of self-efficacy from other likely sources of influence. In this way, it is important for developers to demonstrate that the subscales of the developed instrument measure only sources of influence that they are meant to measure and are not confounded by aspects of the construct measured by another subscale. On the MTEBI instrument, Enochs et al. (2000) identified two subscales for the instrument: personal mathematics teaching efficacy (PMTE) and mathematics teaching outcome expectancy (MTOE). Item analysis was completed for both subscales. Likewise, it is very important to minimize measurement error on research instruments, since the maximum correlation that can be observed between two variables is the product of their reliabilities (Goodwin & Leech, 2006), limiting the relationships that can be detected.

Assemble the Development Team and Plan the Validity Argument

The planning phase is by far the most detailed and involved phase of instrument development, but we argue that it is also the most critical to developers' success. By carefully considering the questions presented in this phase, developers can identify the specific expertise the development team requires and plan in advance to collect the validity evidence necessary to support the instrument's intended uses. Along with these questions, the five sources of validity evidence described in the *Standards* (AERA et al., 2014) provide a useful framework for considering the expertise and evidence developers need. There is, of course, no hard and fast rule governing how much expertise, and in which areas, must be included on the development team, or which forms of validity evidence any given instrument requires. As the *Standards* suggest, "decisions about what type of evidence are important for the validation argument in each instance can be clarified by developing a set of propositions or claims that support the proposed interpretation" (p. 12).

The questions posed in this chapter may serve as a guide to help developers identify the propositions and claims they intend to make and to develop a validity argument supporting them. Further, consideration of these design questions may also help developers determine when the validity evidence needed is beyond the expertise of the current development team, prompting them to seek the assistance of colleagues with the relevant skills. In the sections which follow, we present how some of the claims implied by these questions might be supported with evidence from the sources described in the *Standards*. Note that the following sections are not exhaustive, due to limitations of space. It is incumbent upon instrument developers to critically examine these questions, and to build a case that they have answered them adequately for the instruments' proposed uses. Further, it is incumbent upon instrument users to evaluate developers' claims against the evidence provided to determine whether that case is sound and supported by the evidence, as "validation is the joint responsibility of the test developer and the test user" (AERA et al., 2014, p. 13).

Validity Evidence Based on Test Content

When considering the test content, it may be helpful to have scholars or researchers who study the measured content or teachers who instruct it on the development team. Classroom teachers developing (or selecting) an assessment for use within their own classrooms may want to get comments and feedback from colleagues about how the content is represented, while a curriculum developer producing a similar assessment for widespread use may need to add content experts to the team, conduct a systematic review of literature on the topic, and send drafts of the assessment out for expert review. Validity evidence based on test content includes, but is not limited to, review by a panel of experts, a systematic review of the scholarly literature, construct maps or learning trajectories, and analyses of alignment with relevant content standards (AERA et al., 2014; Sireci & Faulkner-Bond, 2014). In all cases, however, if developers wish to make a claim that their instrument measures specific content, then they must present validity evidence to convince the end user that it does so sufficiently for the intended interpretation and use.

Validity Evidence Based on Response Processes

The key consideration for this source of validity evidence is that test takers are being engaged in response processes that allow the construct of interest to be measured as intended (i.e., avoid the problems of construct irrelevance and/or construct underrepresentation; see AERA et al., 2014, p. 12). For example, the Revised SMPs Protocol is conducted by observing live or video-recorded teaching because, as discussed prior, the other possible response processes (lesson plans or simulated lessons) underrepresent the construct, and thus introduce measurement error for the intended purpose. Validity evidence based on response processes includes, but is not limited to, logical analyses comparing the response processes to the construct of interest, cognitive interviews (aka think-aloud protocols), and focus groups (AERA et al., 2014; Castillo-Díaz & Padilla, 2013; Padilla & Benítez, 2014), and the development team may need to include members with expertise in these techniques. If developers wish to make a claim that their instrument engages test takers in specific types of thinking, analyses, or processes, they must provide reasonable evidence to that effect.

Validity Evidence Based on Internal Structure

This form of evidence deals with the degree to which performance of the items, scales, and subscales of the instrument adhere to the theories on which they were built. Many analyses related to internal structure might be sophisticated enough to require a psychometrician, making it important for developers to consider these needs early enough to add psychometric expertise to the development team, if necessary. Validity evidence based on internal structure includes, but is not limited to, tests of measurement invariance (such as differential item functioning analyses), factor analyses, and measures of internal consistency for subscales (AERA et al., 2014; Rios & Wells, 2014). If the instrument purports to measure components or specific aspects of a construct separately, for example, developers must provide evidence that it does so.

Validity Evidence Based on Relations to Other Variables

Many instruments measure constructs that developers expect to relate in predictable ways to other constructs or variables. This form of evidence, then, concerns whether the scores produced by the instrument are correlated with measures of similar constructs (or related things), and uncorrelated with measures of different constructs (or unrelated things). Validity evidence based on relations to other variables includes convergent evidence, discriminant evidence, predictive evidence, and test–criterion relationships (AERA et al., 2014; Oren, Kennet-Cohen, Turvall, & Allalouf, 2014). When developing an instrument that measures a construct for which other instruments exist, or which is related to other constructs for which instruments already exist, it is often valuable to include these related measures as part of the pilot process to collect this validity evidence.

Validity Evidence Based on Consequences

This source of validity evidence is discussed briefly here as developers should certainly consider the intended and possible unintended consequences that might result from use of their instrument, but this evidence is typically not collected until development is complete and the instrument is in use. Validity evidence based on the related consequences of testing includes, but is not limited to, evidence that use of the instrument is in fact leading to the desired outcomes, evidence of limited or acceptable unintended consequences, and evidence that certain subgroups are not experiencing systematically different consequences from instrument use (AERA et al., 2014; Lane, 2014).

Evidence of Reliability/Precision

The *Standards* (AERA et al., 2014) describe reliability/precision as separate from validity, but with strong implications for validity. If the scores produced by an instrument are neither reliable nor precise, then inferences and decisions based on the instrument will be equally unreliable and imprecise. Thus, developers should plan to collect evidence of reliability/precision during development, as well.

Producing Items and Forms: The Building Blocks of the Instrument

In this second major phase, developers craft items for inclusion in the instrument. Note that we use the term "items" rather broadly, to include traditional test items (e.g., multiple choice, matching, and completion items), as well as Likert-scale items, indicators on an observation protocol, rubric criteria, or any other aspect of an instrument designed to specifically measure a discrete aspect or component of the construct, content, or skill. When constructing items, developers should consult one or more of the numerous resources discussing best practice for the item types used (discussion of specific item design concerns is beyond the scope of this chapter). Developers should construct items according to the construct map, conceptual model, or theoretical framework used during the planning phase. Developers must also attend to the validity evidence they plan to collect during this phase before moving on to the pilot phase.

Producing Items to Formatively Assess Knowledge of Equivalence

When developing or selecting items for formative classroom assessment, it is beneficial to be familiar with common answers to items. In the previous example of 8 + 4 = __ + 5, the development team (which included researchers and teachers) gathered data from 752 elementary school students to learn that the most common answers were: 12, 17, 12 and 17, and 7 (Falkner, Levi, & Carpenter, 1999). The question specifically stated, "What number would you put in the box to make this a true number sentence?" To further understand students' reasoning for each response, extensive cognitive interviews were conducted to provide validity evidence on the test content and response process. This development work and pilot testing (discussed more later) informed a framework on the multiple ways students view the equal sign. For example, students who answered "17," think one should "add it all up" when viewing multiple numbers across an equal sign, rather than viewing the equal sign as "the same as" and answering "7" correctly.

Producing Items to Evaluate the SMPs

The development of the Revised SMPs Protocol drew upon an eight-stage process (see Artino, La Rochelle, Dezee, & Gehlbach, 2014; Smith, Jones, Gilbert, & Wieman, 2013), with some revisions (see Table 2.4). Relevant literature was examined closely to better understand the construct. Experts and potential users were interviewed to explore how such an instrument might be used and what gaps might be filled with such an instrument. Data were qualitatively analyzed to draw conclusions and generate ideas for possible indicators (items). The fourth stage focused on developing indicators for the protocol. Indicators were written that adhered to the themes generated from the inductive analysis at the previous stage and drawn from related instruments and revised to best fit the construct under investigation. At the fifth stage, subject-matter experts examined the protocol's indicators and provided feedback. Validity evidence based on test content was gathered during these first five stages. Feedback from the review panel was incorporated to generate a product to share with potential users and gather data during interviews with potential users. Interview data became validity evidence based on response processes. Stages one through six provided a means to gather content and response processes evidence, which prepared the research team to conduct a pilot study.

Producing Items to Research Self-Efficacy

If researchers want to develop items from scratch, they could conduct interviews and an extensive literature review to determine characteristics of that which is measured. In the case of the MTEBI, Enochs et al. (2000) employed another common approach by adapting items from a similar instrument developed for a different context. The instrument used

Table 2.4 Description of the Eight-Stage Process for Validating Observation Protocols

Stage	Description of Stage	Actions Completed During This Study
1	Literature review	Examined initial protocol; reviewed literature on observation protocols and mathematical proficiency.
2	Conducting interviews with content experts and potential users of tool	Conducted interviews with expert panel consisting of K-12 math teachers, math coaches, curriculum coordinators, mathematicians, and mathematics teacher educators.
3	Synthesizing data from literature review and interviews	Employed inductive analysis to generate themes that should be found in indicators.
4	Item development	Revised indicators from related instruments and added new indicators based upon thematic analysis.
5	Expert panel validation	Submitted revised look-for protocol to expert panel.
6	Conducting interviews with potential users of tool and synthesizing from these data	Conducted one-on-one and small-group interviews with K-12 math teachers, a principal, curriculum coach, mathematicians performing observations in schools, and mathematics educators.
7	Pilot testing tool	Performed 288 observations with the revised look-for protocol.
8	Conducting psychometric analysis of data from use of the tool	Performed reliability analysis using Cronbach's alpha (internal consistency) and test–retest reliability.

five-point Likert-scale items (ranging from *Strongly Agree* to *Strongly Disagree*) to measure two subscales: PMTE and MTOE. Researchers had to ensure that the items only reflected one of the subscales and not both (Riggs & Enochs, 1989). Example items included: "[PMTE]: Even if I try very hard, I will not teach mathematics as well as I will most subjects; [MTOE]: The mathematics achievements of some students cannot generally be blamed on their teachers" (Enochs et al., 2000, p. 195). Item tryouts, administered to individuals representative of the target population, helped researchers understand psychometric properties of the items, specifically item difficulty and if the instrument was able to distinguish among individuals at different levels (AERA et al., 2014). Conducting cognitive interviews can reduce threats to validity from self-report items (Biemer, Groves, Lyberg, Mathiowetz, & Sudman, 1991) by allowing researchers to uncover reasons individuals respond to items in certain ways or to determine which items are misleading or incomplete for the given construct (Desimone & Le Floch, 2004).

Pilot Testing

During this phase, the goal is to get a small sample of participants who are representative of the population of interest to take the assessment. The pilot sample may be randomly or purposively selected, depending on the purpose of the instrument. Similarly, the sample size and format of the pilot may depend on the proposed interpretation and use of the instrument and the validity evidence that developers intend to collect. Qualitative feedback from test takers about items' format and difficulty could be obtained from a very small sample, while some psychometric analyses might require a much larger sample to yield informative results.

Results from pilot administrations should be examined to identify and correct possible sources of measurement error. Thus, items may be flagged for further investigation because they perform differently than other items meant to measure the same thing, are inconsistent with data obtained from other sources of information about the content or construct measured, or because they appear to assess certain groups of test takers differently when there is no reason to suspect that these groups perform differently on the content or construct measured. In such cases, items may need to be adjusted and repiloted or discarded in favor of other items which perform as intended. The piloting process is also designed to test potential issues that were raised in the previous two stages, collecting evidence that these concerns have either been fully resolved, or that they remain a problem and require further attention to resolve them.

Piloting Classroom Assessments

When items and instruments are developed for classroom assessment purposes, the pilot testing phase should include a representative sample from the grade levels in which the mathematical content applies. In our earlier addition example of $8 + 4 = \_\_ + 5$, the computation is accessible for all grade levels, because the purpose is to focus on equality and not include yet another content or related variable. In their use of the open number sentences to develop a module on early algebraic thinking, Fisher, Thomas, Jong, Schack, and Dueber (2019) created the following item that was similar to use with second-grade students: $10 + 10 = \_ + 5$. As a result, students' answers reflected the same ideas about the equal sign along with a new concept about their operational understanding. One student answered "4," reasoning that "5 four times is 20," which would be correct if the plus sign was a multiplication sign. The cause of the student's error could only be determined by a cognitive interview or a small group pilot administration in which researchers discuss responses with students, however, illustrating the need for planned validity evidence to shape the pilot.

Piloting Evaluation Tools

Bostic and colleagues (2019) carried out stage seven for the Revised SMPs Protocol: pilot testing with a large enough group to gather evidence for internal structure and relationships to other variables. While there is a concern for using a large sample at this stage, it was desirable because (1) factor analyses typically require relatively large samples (Mundfrom, Shaw, & Ke, 2005) and (2) investigating relationships between the outcome variable and demographic variables (e.g., school type, location of school [rural, suburban, and rural], and teacher's sex) requires more administrations to decrease potential variance in the sample. At this stage, 288 observations were made using the protocol, coded with both live and video-recorded data. Half of the observations took place before a professional development program focused on the SMPs, while the other half occurred after 100 hours of professional development. The size of the sample and the schedule of administrations allowed developers to collect the planned validity evidence during piloting.

Observation protocols, similar to other instruments that must be scored by human raters, should address rater training and inter-rater reliability as part of their development and initial validation. The piloting phase presents a natural opportunity to address these concerns. Raters, who demonstrated sufficient understanding of the SMPs before being selected, were trained during an eight-hour program. During this program, raters watched one video with the development team, identified potential indicators, and then discussed their findings as a group. Disagreement across raters was negotiated to the point where everyone agreed on the same indicators. Next, raters independently coded a second video and submitted their codes for inter-rater agreement. Inter-rater agreement expresses the degree to which raters' codes are identical, whereas inter-rater reliability expresses the similarity of raters' coding patterns (Crocker & Algina, 2006). Data collected during the pilot and during rater training provided sufficient validity evidence based on internal structure and relations to other variables, as well as evidence of inter-rater reliability and internal consistency. Interested readers should consult Bostic et al. (2019) for further discussion and for the specific validity evidence collected.

Piloting Research Instruments

After developing items for a research instrument on self-efficacy, and conducting cognitive interviews, it is necessary to conduct a pilot test of the full instrument in its (expected) final form. Researchers should locate a sample of participants that represents the target population that is large enough to support the psychometric analyses they intend to use. Factor analyses (whether exploratory or confirmatory) are among the most common and informative analyses to run during instrument development. While research into the minimum sample sizes required for factor analyses has not found broad consensus, it has revealed that many factors influence the sample size necessary to recover stable factors in any sample (MacCallum, Widaman, Zhang, & Hong, 1999). If researchers wish to generate item response theory (IRT) parameters for items, or run differential item functioning (DIF) analyses, much larger response samples are required. Thus, decisions about the pilot sample are best made in consultation with someone familiar with these methods. In addition, unless an established, strong theoretical framework was used to carefully craft items that measure subconstructs without overlap, exploratory factor analysis (EFA) should be used before confirmatory factor analysis (CFA) is defensible (Bandalos & Finney, 2010). In either case, however, factor loadings for individual items, as well as internal consistency analyses, can help identify items that measure the intended construct(s) best, and which items underperform. Weaker items may then be dropped or modified to improve their performance. Enochs and colleagues (2000) provide detailed evidence of the validation for the MTEBI and interested readers are referred to their paper.

Full Implementation

Developers should compare all of the validity evidence collected through the planning, production, and piloting phases to the seven questions of the summary statement considered at the beginning of this process. If developers have sufficient evidence to support their argument that the instrument is valid for its intended purposes, they may conclude that it is ready for use. Once the instrument developed is ready for full implementation, it does not mean that questions related to validity and validation are put to rest. The developer and/or the test user must continue to monitor use of the instrument in applied settings to ensure that decisions made on its basis continue to be sound. Changes over time in the intended context or target population for which the instrument was designed may lead to changes in how the instrument performs, requiring new investigations to determine if it remains valid for the originally intended purposes. Also, as theory around the construct of interest continues to develop and evolve, researchers and evaluators should consider how these developments might affect the intended uses of their instruments. Across all measurement contexts and intended uses, however, it is only during full implementation of an instrument that the consequences of its use can be fully understood. Thus, both developers and users should continually consider the questions shown in Table 2.5. While validation is an ongoing process, it is not intended to be a never-ending process (AERA et al., 2014; Kane, 2013). Answers to these final questions about consequences of test use, however, may reopen a validation effort that was previously thought closed.

Table 2.5 Development Questions to Consider During the Full Implementation

Questions	Implications for Development/Validity Evidence
a. What evidence is there to suggest that the instrument is supporting its intended outcomes and consequences?	• Evidence that use of the instrument achieves its intended outcomes helps justify its continued use (validity evidence based on related consequences).
b. What unintended benefits may have come from full implementation and use of the instrument?	• Evidence of positive unintended outcomes may further justify its continued use and provide implications for future research (validity evidence based on related consequences).
c. What unintended negative consequences may have come from full implementation and use of the instrument for test users, test takers, other stakeholders, or for vulnerable subgroups of any of these?	• Evidence of negative unintended consequences suggests that a cost-benefit analyses be conducted to determine whether use of the instrument should be maintained, adjusted, or discontinued (validity evidence based on related consequences and implications for fairness).

Conclusion

The purpose of this chapter is to provide possible design questions that developers may use as a resource when planning, producing, and piloting an instrument. The development questions offered here are not meant to be exhaustive and may not fit all possible instruments developed or all the possible purposes for which readers may develop them. The questions and examples discussed in this chapter are meant, however, to capture many common development and validation concerns that are likely to arise when developing instruments for assessment, evaluation, and research in a mathematics education context. Instrument development and validation is complex, detailed, and nuanced work. By carefully and systematically attending to validity concerns throughout the process, however, developers can produce instruments that work well for their intended purposes.

Author Note

The authors acknowledge Michele Carney for meaningful contributions to several team discussions that shaped this chapter, and for providing substantive comments on an early draft.

References

Alonzo, A. C. (2007). Challenges of simultaneously defining and measuring knowledge for teaching. *Measurement: Interdisciplinary Research and Perspectives*, *5*(2–3), 131–137. doi:10.1080/15366360701487203

American Educational Research Association, American Psychological Association, & National Council on Measurement in Education. (2014). *Standards for educational and psychological testing.* Washington, DC: American Educational Research Association.

Anderson, L. W., & Krathwohl, D. (Eds.). (2001). *A taxonomy for learning, teaching, and assessing: A revision of Bloom's taxonomy of educational objectives.* New York: Longman.

Artino, A., Jr., La Rochelle, J., Dezee, K., & Gehlbach, H. (2014). Developing questionnaires for educational research: AMEE guide no 87. *Medical Teacher, 36,* 463–474.

Bandalos, D. L., & Finney, S. J. (2010). Factor analysis. In G. R. Hancock & R. O. Mueller (Eds.), *The reviewer's guide to quantitative methods in the social sciences.* New York, NY: Taylor & Francis.

Bandura, A. (1977). Self-efficacy: Toward a unifying theory of behavioral change. *Psychological Review, 84*(2), 191–215. doi:10.1037/0033-295x.84.2.191

Biemer, P. P., Groves, R. M., Lyberg, L. E., Mathiowetz, N. A., & Sudman, S. (1991). *Measurement errors in surveys.* New York, NY: Wiley.

Black, P., Wilson, M., & Shih-Ying, Y. (2011). Road maps for learning: A guide to the navigation of learning progressions. *Measurement: Interdisciplinary Research and Perspectives, 9*(2/3), 71–123. doi:10.1080/15366367.2011.591654

Bloom, B. S. (Ed.). (1956). *Taxonomy of educational objectives: The classification of educational goals.* New York: Longmans, Green.

Bostic, J., Krupa, E., & Shih, J. (2019). *Quantitative measures of mathematical knowledge: Researching instruments and perspectives.* New York, NY: Routledge.

Bostic, J., & Matney, G. (2013). Overcoming a common storm: Designing PD for teachers implementing the common core. *Ohio Journal of School Mathematics, 67,* 12–19.

Bostic, J., & Matney, G. (2016). Leveraging modeling with mathematics-focused instruction to promote other standards for mathematical practice. *Journal of Mathematics Education Leadership, 17*(2), 21–33.

Bostic, J., Matney, G., & Sondergeld, T. (in press). A lens on teachers' promotion of the Standards for Mathematical Practice. *Investigations in Mathematics Learning, 11*(1), 69–82. doi:10.1080/19477503.2017.1379894

Briggs, D. C. (2004). Comment: Making an argument for design validity before interpretive validity. *Measurement: Interdisciplinary Research and Perspectives, 2*(3), 171–191. doi:10.1207/s15366359mea0203_2

Brookhart, S. M., Guskey, T. R., Bowers, A. J., McMillan, J. H., Smith, J. K., Smith, L. F., . . . Welsh, M. E. (2016). A century of grading research: Meaning and value in the most common educational measure. *Review of Educational Research, 86*(4), 803–848. doi:10.3102/0034654316672069

Carney, M., Totorica, T., Cavey, L., & Lowenthal, P. (2019). Developing construct maps for teacher candidate attentiveness. In J. Bostic, E. Krupa, & J. Shih (Eds.), *Quantitative measures of mathematical knowledge: Researching instruments and perspectives.* Abingdon, UK: Routledge.

Carpenter, T. P., Franke, M. L., & Levi, L. (2003). *Thinking mathematically: Integrating arithmetic and algebra in elementary school.* Portsmouth, NH: Heinemann.

Castillo-Díaz, M., & Padilla, J.-L. (2013). How cognitive interviewing can provide validity evidence of the response processes to scale items. *Social Indicators Research, 114*(3), 963–975. doi:10.1007/s11205-012-0184-8

Crocker, L., & Algina, J. (2006). *Introduction to classical and modern test theory* (2nd ed.). Mason, OH: Wadsworth Publishing.

Desimone, L. M., & Le Floch, K. C. (2004). Are we asking the right questions? Using cognitive interviews to improve surveys in education research. *Educational Evaluation and Policy Analysis, 26*(1), 1–22.

Enochs, L. G., Smith, P. L., & Huinker, D. (2000). Establishing factorial validity of the mathematics teaching efficacy beliefs instrument. *School Science and Mathematics, 100*(4), 194–202.

Falkner, K. P., Levi, L., & Carpenter, T. P. (1999). Children's understanding of equality: A foundation for algebra. *Teaching Children Mathematics, 6*(4), 232.

Fisher, M. H., Thomas, J., Jong, C., Schack, E. O., & Dueber, D. (2019). Comparing preservice teachers' professional noticing skills in elementary mathematics classrooms. *School Science and Mathematics, 119*(3), 142–149.

Gerber, D., Bostic, J. D., & Lavery, M. R., (2018, October). *Promoting understanding and applications of validity for teachers' classroom assessment.* Paper presented at the National Council for Measurement in Education Classroom Assessment Conference, Lawrence, KS.

Gerber, D., Lavery, M., & Bostic, J. (in press). Making valid instructional decisions: Teaching educators to consider validity evidence. In S. L. Nichols & D. Varier (Eds.), *Teaching on Assessment.* Charlotte, NC: Information Age Publishers.

Goodwin, L. D., & Leech, N. L. (2006). Understanding correlation: Factors that affect the size of *r. Journal of Experimental Education, 74*(3), 251–266.

Haertel, E. H., & Herman, J. L. (2005). A historical perspective on validity arguments for accountability testing. *Yearbook of the National Society for the Study of Education, 104*(2), 1–34. doi:10.1111/j.1744-7984.2005.00023.x

Kane, M. T. (2004). Certification testing as an illustration of argument-based validation. *Measurement: Interdisciplinary Research and Perspectives, 2*(3), 135–170. doi:10.1207/s15366359mea0203_1

Kane, M. T. (2013). Validating the interpretations and uses of test scores. *Journal of Educational Measurement, 50*(1), 1–73. doi:10.2307/23353796

Kane, M. T. (2016). Validation strategies: Delineating and validating proposed interpretations and uses of test scores. In S. Lane, M. R. Raymond, & T. M. Haladyna (Eds.), *Handbook of test development* (2nd ed., pp. 64–80). New York, NY: Routledge & Taylor & Francis Group.

Kilpatrick, J., Swafford, J., & Findell, B. (2001). *Adding it up: Helping children learn mathematics.* Washington, DC: National Academy Press.

Krathwohl, D. R. (2002). A revision of Bloom's taxonomy: An overview. *Theory into Practice, 41*(4), 212–218. doi:10.1207/s15430421tip4104_2

Lane, S. (2014). Validity evidence based on testing consequences. *Psicothema, 26*(1), 127–135.

MacCallum, R. C., Widaman, K. F., Zhang, S., & Hong, S. (1999). Sample size in factor analysis. *Psychological Methods, 4*(1), 84–99. doi:10.1037/1082-989X.4.1.84

Maloney, A. P., Confrey, J., & Nguyen, K. H. (Eds.). (2014). *Learning over time: Learning trajectories in mathematics education.* Charlotte, NC: Information Age Publishing.

Matney, G., Bostic, J., & Lavery, M. (2019). A validation process for complex pedagogical knowledge: The standards for mathematical practice knowledge assessment. In J. Bostic, E. Krupa, & J. Shih (Eds.), *Quantitative measures of mathematical knowledge.* New York, NY: Routledge.

Messick, S. (1989). Validity. In R. L. Linn (Ed.), *Educational measurement* (3rd ed., pp. 13–103). New York, NY: Macmillan Publishing Co, Inc.

Mislevy, R. J., & Riconscente, M. M. (2006). Evidence-centered assessment design. In S. M. Downing & T. M. Haladyna (Eds.), *Handbook of test development* (pp. 61–90). Mahwah, NJ: Erlbaum.

Mundfrom, D. J., Shaw, D. G., & Ke, T. L. (2005). Minimum sample size recommendations for conducting factor analyses. *International Journal of Testing*, 5(2), 159–168. doi:10.1207/s15327574ijt0502_4

National Council of Teachers of Mathematics. (2000). *Principles and standards for school mathematics*. Reston, VA: Author.

Oren, C., Kennet-Cohen, T., Turvall, E., & Allalouf, A. (2014). Demonstrating the validity of three general scores of PET in predicting higher education achievement in Israel. *Psicothema*, 26(1), 117–126. doi:10.7334/psicothema2013.257

Padilla, J.-L., & Benítez, I. (2014). Validity evidence based on response processes. *Psicothema*, 26(1), 136–144.

Pellegrino, J., DiBellow, L., & Goldman, S. (2016). A framework for conceptualizing and evaluating the validity of instructionally relevant assessments. *Educational Psychologist*, 51(1), 59–81.

Riggs, I., & Enochs, L. (1989, March). *Toward the development of an elementary teacher's science teaching efficacy belief instrument*. Paper presented at the annual meeting, of the National Association for Research in Science Teaching, San Francisco.

Rios, J., & Wells, C. (2014). Validity evidence based on internal structure. *Psicothema*, 26(1), 108–116.

Schilling, S. G. (2004). Conceptualizing the validity argument: An alternative approach. *Measurement: Interdisciplinary Research and Perspectives*, 2(3), 178–182.

Sireci, S., & Faulkner-Bond, M. (2014). Validity evidence based on test content. *Psicothema*, 26(1), 100–107.

Smith, M., Jones, F., Gilbert, S., & Wieman, C. (2013). The classroom observation protocol for undergraduate STEM (COPUS): A new instrument to characterize university STEM classroom practices. *CBE-Life Sciences Education*, 12, 618–627.

Wilson, M. R. (2005). *Constructing measures: An item response modeling approach*. Mahwah, NJ: Lawrence Erlbaum Associates.

Wilson, M. R. & Wilmot, D. (2019 [this volume]). Gathering validity evidence using the Bear Assessment System (BAS): A mathematics assessment perspective. In J. Bostic, E. Krupa, & J. Shih (Eds.), *Assessment in mathematics education contexts: Theoretical frameworks and new directions*. New York, NY: Routledge.

3 Measure Validation as a Research Methodology for Mathematics Education

Erik Jacobson and Rebecca Borowski

In the final two decades of the 20th century, mathematics education researchers "erected a monument to qualitative research methods and non-experimental modes of inquiry" (Silver, 2004, p. 154). This focus significantly advanced the field of mathematics education by detailing the important role of context in studying educational phenomena. In the first two decades of the 21st century, some interest in quantitative methods in mathematics education research has returned, driven in part by new requirements from funding agencies and changes in education policy. For example, at the national level, the No Child Left Behind Act of 2001 has had a large impact on schooling by defining student, teacher, and school accountability in terms of standardized tests. Concurrently, as the field of mathematics education has advanced, existing measures are no longer appropriate or have the wrong focus for questions of current interest. These two streams of change have converged: Mathematics education studies are increasingly involving quantitative analyses that rely on educational and psychological measures and at the same time mathematics educators are increasingly called upon to design new instruments to measure the phenomena they study. This confluence suggests it is timely to reconsider how an increased focus on measure design and validation can advance the field of mathematics education.

Many important findings that have accrued in mathematics education have not yet been studied systematically across contexts. By operationalizing constructs, measures provide an important means to study phenomena across contexts and at scale. Although the field of mathematics education values theory, and many theoretical frameworks and approaches to research in mathematics education are empirically supported by case studies, interviews, and other forms of qualitative research, much of the theory in the field involves phenomena that could be but have not yet been operationalized with a psychometric measure. The measure development process demands a different kind of articulation of theory than research reports of qualitative data do, and the validation process often surfaces important, unexamined questions. Successful validation studies benefit the field not only by increasing confidence in a specific measure for a specific use, but also by confirming the generality and explanatory power of the relevant theory. Unsuccessful validation studies weaken validity arguments for specific uses of measures but also advance the field by highlighting critical gaps in theory, illuminating the difficulties in measuring some constructs, and suggesting new research questions.

In this chapter, we argue that measure validation is not a routine exercise but rather that it has important, underused potential as research methodology to advance the field of mathematics education. First, we describe measure development and current scholarly norms for validation studies. Then we present two examples to illustrate our thesis that validation studies are an important form of inquiry in mathematics education. In the first case, a measure was developed to measure the targeted constructs and the validation studies supported the intended use. In the second case, we describe a validation study that did not support the intended use of the measure. We argue that both cases led to important research findings with implications for the field. We conclude the chapter by discussing the need for dissemination outlets that will publish methodologically sound validation studies, regardless of their outcomes.

In the first case, we examine the validity of a measure of preservice teachers' knowledge and motivation for teaching multidigit arithmetic. We summarize several validation studies that both lend credence to the intended interpretation of scores on the measure in the context of teacher education and importantly advanced the field by raising new questions for research. The second case examined the validity of a measure of fifth-grade students' knowledge of the number line representation. After synthesizing a comprehensive framework based on prior research and writing items aligned with the framework, a think-aloud item response interview study revealed that the cognition elicited by the items was far more varied than the scope the original framework allowed. Thus, because responses on the items did not support the intended inferences about student knowledge, the study advanced the field by revealing previously unnoticed theoretical gaps and critical questions for future research.

Measure Validation

Instrument validation is the two-part process of (1) developing an interpretive argument that details the context of use and how scores will be interpreted in the context and (2) a validity argument that uses evidence to evaluate the interpretive argument (Kane, 2004, 2013). According to the *Standards for Educational and Psychological Testing* (American Educational Research Association [AERA], American Psychological Association [APA], & National Council on Measurement in Education [NCME], 2014), *validity* refers to "the degree to which all the accumulated evidence supports the intended interpretation of test scores for the proposed use" (p. 14). Thus, the focus of validation is specific interpretations of a test score for a specific use; not the test or instrument itself in the absence of context. In this section, we compare two approaches to validation, and argue that one approach better foregrounds the role of theory.

The first approach to validation is described in the *Standards for Educational and Psychological Testing* (AERA et al., 2014). The validation argument can include five categories of validity evidence: test content, response process, internal structure, relations to other variables, and consequences of testing. Expert judgments often provide evidence about whether the instrument content represents the construct the instrument is intended to measure. Evaluating the response process (e.g., think-aloud item response interviews; Peterson, Peterson, & Powell, 2017) "can provide evidence concerning the fit between the construct and the detailed nature of performance or response actually engaged in by examinees" (AERA et al., 2014, p. 12). Psychometric evidence is often used to evaluate how items on a test relate to one another and the overall score. Correlation evidence is often examined in the fourth category to determine whether these relationships are aligned with theoretical predictions. The main question for examinations of the consequences of testing, the fifth and final category, is whether tests are fair and have equitable outcomes for users.

The structure for validation arguments described in the *Standards* (AERA et al., 2014) can help clarify validation efforts by making sure validation arguments address each of the five categories of evidence. However, because each category of evidence can be addressed by many different kinds of studies, developers who use this structure risk bias toward more convenient sources of evidence rather than evidence that is most relevant. Reviewers may be persuaded by validation arguments that address all the categories of evidence, overlooking weak links in the argument. Especially for less familiar reviewers and novice researchers, the *Standards* approach to validation may minimize the key point that validation is a persuasive argument, not a checklist. Another limitation to the *Standards* approach is the implicit assumption of mature theory for the construct being measured. For example, the test content category of evidence presumes there is a well-defined construct and the relations between variables category presumes such relationships are already specified by extant theory. Yet theory defining and relating constructs in mathematics education is frequently not yet fully developed because the field is so new. In several of the main areas of active research, theory and instruments are developing together.

A prominent example of mathematics education instrumentation and theory developing together is the Learning Mathematics for Teaching instruments (LMT, n.d.; Schilling, 2004, 2007). Schilling approached instrument validation in a way that we argue is well-suited for cases where the theory of the construct is not yet well-developed. The approach begins by explicitly naming the underlying theoretical assumptions and then organizes empirical evidence to address three complementary domains of validation. Researchers begin by articulating the theoretical assumptions that provide a foundation for the construct and instrument design; then they proceed by identifying empirically falsifiable inferences that can be examined in validation studies. Three categories of theoretical assumptions and inferences are considered. *Elemental* assumptions and inferences focus on the test questions, and this category intersects both *test content* and *response process*, two validity categories described in the *Standards* (AERA et al., 2014). *Structural* assumptions and inferences deal with constructs and the relation of constructs with questions. This category is similar to the *internal structural* validity category described in the *Standards*. Schilling's third category includes *ecological* assumptions and inferences, and this category intersects both the *relations with other variables* and *test consequences* categories in the *Standards*. We found Schilling's approach useful because it both addressed the main categories of the *Standards* and is well aligned with theory development. Moreover, Schilling's approach provides more structure for the argument aspect of validation, not just the content. For these reasons, we used Schilling's framing of validation arguments in both of the cases we present in this chapter.

Case 1: Measuring Integrated Knowledge and Motivation for Teaching Multidigit Arithmetic

Some measures developed for teachers have focused on knowledge and others have focused on affective or belief-related domains. For example, motivation is a construct that involves feelings of mathematical anxiety as well as beliefs about the value of mathematics and self-concept of ability, and these constructs are theoretically and empirically distinct from mathematical knowledge (Newton, 2008, 2009).

Although the same teacher preparation experiences are thought to shape both knowledge and beliefs, existing measures for these constructs are not well aligned. In the case of mathematics teacher education, for example, different measures often define mathematics in a different way, with some measures focused on a particular mathematical domain, such as rational numbers, whereas other measures focus on broad areas of mathematics, such as all of the mathematical content taught in the elementary curriculum. The lack of alignment between measures makes it difficult to understand the relative impact of teacher preparation on knowledge and beliefs because differences between constructs are conflated with differences in the mathematics at stake.

The first case we discuss involved a measure for elementary preservice teachers focused on the mathematical topic of multidigit arithmetic. The instrument we developed used a novel approach to integrate the operationalization of the knowledge and motivation constructs at the item level. This strategy is promising for understanding the differential impact of teacher preparation across multiple outcomes. The intended use of the measure was to provide feedback on the design of a teacher preparation course and an associated field experience.

Theoretical Construct and Measure Design

The Mathematical Proficiency for Teaching (MPT; Jacobson, 2013, 2017a) framework organizes the multiple goals of mathematics teacher education into a coherent focus for research. It includes knowledge constructs such as Pedagogical Content Knowledge (PCK; Shulman, 1986) and constructs of productive disposition for teaching such as mathematics-related emotions, attitudes, and beliefs that have empirical relationships with student learning (Jacobson & Kilpatrick, 2015). Teacher education research has often focused on single outcomes (e.g., pedagogical content knowledge for mathematics, Baumert et al., 2010; motivation beliefs for fractions, Newton, 2009). The MPT framework supports research on multiple outcomes and potential interactions.

We designed a new instrument to study the relationships between different MPT constructs, and we focused on a subset of elementary school mathematics content—multidigit addition and subtraction—and a selection of MPT constructs. We chose to operationalize knowledge and motivation constructs because research suggests these constructs are positively correlated with student learning, yet the constructs are theoretically distant from each other and therefore afford increased contrast. A practical question we aimed to answer was how elementary teacher mathematics methods classes shaped the MPT of preservice teachers.

The knowledge construct we measured is PCK (e.g., Cai, 2005; Schmidt, Houang, & Cogan, 2011). Two components of PCK are (1) selecting and using instructional representations and (2) appraising students' conceptions and reasoning. Teachers' mathematics PCK has been positively correlated with students' mathematics achievement in several studies. For example, Hill, Rowan, and Ball (2005) found that student achievement had a similar positive correlation with PCK as it did with socioeconomic status. Another study (Baumert et al., 2010) found that the students of teachers with high PCK scores learned more than the students of teachers with low PCK scores.

Our operationalization of the motivation construct includes three component beliefs—anxiety, self-concept of ability, and value—drawn from the expectancy-value theory of achievement motivation (e.g., Wigfield & Eccles, 2000). Although beliefs can be fixed by adulthood and resistant to change, there is evidence that undergraduate preservice teacher education can lead to changes in motivation beliefs for teaching mathematics (Newton, 2009). Teacher motivation and knowledge both interact to influence practice. Jacobson and Izsák (2015) reported that the positive correlation between teacher's knowledge of instructional representations and self-reported use of these representations in the classroom is completely mediated by motivation for using representations. Thus, we selected MPT constructs that are reasonable goals for elementary mathematics methods class and likely to indirectly influence student achievement.

We wrote multiple choice PCK items after reviewing the research literature and studying published examples. We balanced items between (1) selecting and using instructional representations (Figure 3.1a) and (2) appraising students' conceptions and reasoning (Figure 3.1b); the two kinds of PCK Shulman described. Survey instruments designed to measure beliefs about mathematics use a common phrase across items to designate the mathematical scope. The question, "How good *at math* are you?" (emphasis added; Wigfield & Meece, 1988) measures self-concept of ability at a wide scope (the academic discipline of mathematics), whereas rephrasing the question to "How good *at fractions* are you?" (emphasis added; Newton, 2009) narrows the mathematical scope to the topic of fractions. For our research goals, we wanted to make sure the knowledge and motivation constructs were operationalized within the same mathematical scope. Rather than a common phrase, we used a PCK item to designate the mathematical scope. To measure motivation beliefs (anxiety, self-concept of ability, and value), we paired each PCK item with a set of rating questions (Strongly disagree to Strongly agree; see Figure 3.1c). Thus, the PCK items on the instrument both assess knowledge and define the mathematical scope for the motivation items, and the operationalization of the PCK and motivation constructs in the MPT instrument are mathematically aligned at the item level.

Assumptions, Inferences, and Validation Studies

Following Schilling's (2004, 2007) approach to validity arguments, we considered elemental, structural, and ecological assumptions based on the underlying theoretical framework. For each assumption, we identified inferences—potential consequences of the assumptions that were empirically testable—and then we conducted validation studies for each inference. The assumptions and inferences are listed below (see also Jacobson and Svetina, 2019).

1. Elemental assumption: The items reflect preservice teachers' pedagogical content knowledge and motivation for specific hypothetical teaching situations.

    a. Inference: Preservice teachers' reasons for selecting responses to items will correspond with the planned interpretation of those item responses.

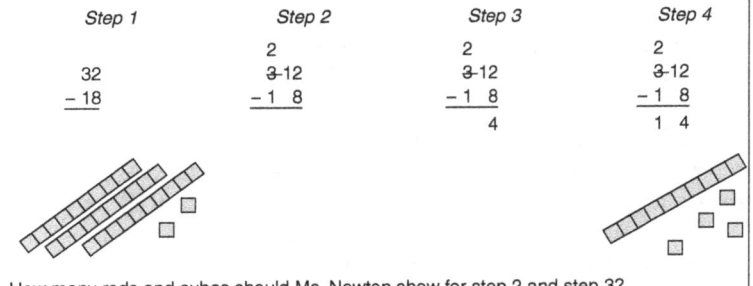

16. Ms. Newton was using cubes that snap together to explain how to record the solution for the problem 32 − 18 = 14. She wrote down four steps, starting with 32 cubes and ending up with 14 cubes.

How many rods and cubes should Ms. Newton show for step 2 and step 3?

a

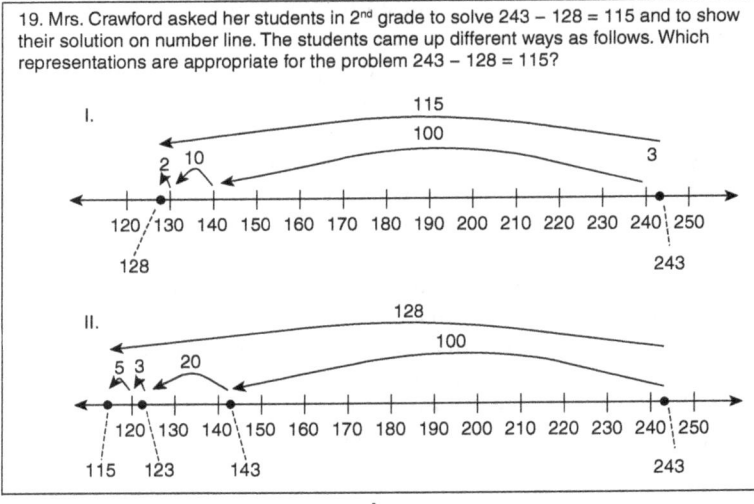

19. Mrs. Crawford asked her students in 2nd grade to solve 243 − 128 = 115 and to show their solution on number line. The students came up different ways as follows. Which representations are appropriate for the problem 243 − 128 = 115?

b

Please CIRCLE ONE OPTION to rate your agreement with the following statements:	Strongly Disagree		Disagree	Agree		Strongly Agree
I am good at answering questions like this one.	1 2	3	4	5	6 7	
I often feel nervous when I try to answer questions like this one.	1 2	3	4	5	6 7	
Elementary teachers should know how to answer this question.	1 2	3	4	5	6 7	
If I try hard, I can usually figure out questions like this one.	1 2	3	4	5	6 7	

c

Figure 3.1 Sample PCK items (a, b) from the MPT survey measure that illustrate the content area of multidigit addition and subtraction. The motivation and teaching self-efficacy survey questions (c) assessed self-concept of ability, anxiety, value, and teaching self-efficacy, and they used PCK questions to describe the mathematical scope of the construct. Figure by Jacobson (2017b), available online at https://doi.org/10.6084/m9.figshare.4793437.v1 under a CC-BY4.0 license.

    b.   Inference: Experts will judge that items provide useful information about preservice teachers' knowledge and motivation for teaching multidigit addition and subtraction.

2.   Structural assumption: The subscales of the instrument which measure different cognitive and noncognitive constructs (i.e., pedagogical content knowledge, anxiety, value, and self-concept of ability) both (1) can be distinguished from each other in the same hypothetical teaching situations and (2) are stable across a set of hypothetical teaching situations related to the mathematical topic of multidigit arithmetic.

    a.   Inference: The raw item scores for each scale will be related to the average raw score, the set of items for each scale will exhibit acceptable reliability, and the average raw scores for the constructs will be related to each other in expected ways but not be the same.

    b.   Inference: Psychometric models based on item response theory and confirmatory factor analysis will not disconfirm that items are more strongly correlated within construct than between construct, and that latent factors are related but distinct.

3.   Ecological assumption: The instruments will capture characteristics of preservice teachers' psychology that can change during a methods course.

    a.   Inference: Preservice teachers who have taken a methods class will have scores reflecting greater mathematical proficiency for teaching than those who have not yet taken a methods class.

    b.   Inference: The instrument will be sensitive to changes in the constructs it was designed to measure such that preservice teachers will have different mathematical proficiency for teaching scores after taking a methods class as compared with their scores before taking the methods class.

We conducted one or more validation studies to test each inference. A complete description of all the validation studies is beyond the scope of this chapter. In what follows we summarize each study briefly, and conclude the section by discussing a validation study for Inference 3.b in more detail. This example illustrates our argument that validation studies have important contributions to theory.

We evaluated Inference 1.a in light of think-aloud item response interviews with 15 preservice teachers. In these interviews, preservice teachers narrated their response process and rationale for each item (i.e., "think aloud"), and researchers asked clarifying questions when necessary but did not provide tutoring or feedback. By comparing preservice teachers' reasoning and the choices they selected on each question, we judged that the intended interpretations for specific choices were justified. To evaluate Inference 1.b, we sent the survey to eight individuals with content expertise and experience teaching elementary preservice teachers. Each of the experts provided ratings and textual feedback on the items. The expert feedback confirmed that the items were appropriate for the intended population and reasonable format and wording for our goals.

We evaluated Inference 2.a by examining the raw scores on the survey of a sample of 119 preservice teachers. Each construct demonstrated reasonable reliability, and the correlations between sum scores on each construct were statistically significant and small, suggesting the constructs were related in the expected ways yet also distinct. For example, the anxiety scale was negatively correlated with all other scales in the measure. We evaluated Inference 2.b by using confirmatory factor analysis and item response theory and collecting additional data for a total sample of 326. The IRT model fit and displayed appropriate psychometric characteristics; the CFA models also fit the data and provided further evidence that the scales were related yet distinct. We evaluated Inference 3.a by using MANOVA techniques to compare the scores of PSTs who had taken the methods course ($n = 54$) with those who had not yet taken the methods class ($n = 65$). We found statistically significant difference between these groups on the knowledge and anxiety scales, but not on the other scales.

We conclude this section by describing the study used to evaluate Inference 3.b in more detail. In this study, we used paired t-tests to compare scores before the methods course with scores after the methods course for a sample of 33 preservice teachers (Figure 3.2). In this design, each preservice teacher acted as her or his own control. We found statistically significant increases in knowledge and in teaching self-efficacy beliefs. However, although anxiety scores decreased and self-concept of ability scores increased as predicted, none of the changes in motivation

Table 3.1 Paired *t*-Tests Showing Statistically Significant Change in Knowledge and Self-Efficacy Beliefs, But Not in Motivation Beliefs

	Before Methods Class		After Methods Class		Difference
	M	SD	M	SD	t ($df = 32$)
Knowledge	−0.32	0.76	0.07	0.65	2.70*
Self-concept of ability	4.97	0.94	5.27	1.08	1.90[a]
Anxiety	3.70	1.32	3.58	1.50	−0.43
Value	6.17	0.69	6.18	0.77	0.08

[a]$p = .066$, * $p < 0.01$

Source: Jacobson, Aydeniz, Creager, Diaga, & Uzan, 2018

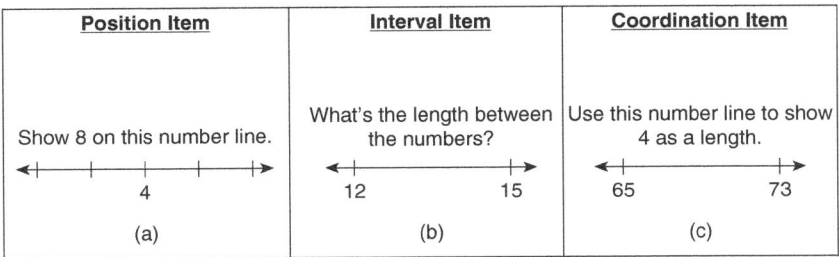

Figure 3.2 Types of items: position (a), interval (b), and coordination of position and interval (c).

beliefs were statistically significant at the 0.05 level. Moreover, the *value* construct showed essentially no change. One contributing factor is that initial levels of the value score had a mean of 6.17 on a 7-point scale. Because the initial values were so high, there was little room on the scale for PSTs to indicate increases in the value belief. We explore the implications of these results in the next section.

Theoretical Implications of the Validation Studies

In this section, we discuss the theoretical implications of the validation studies for the measure of mathematical proficiency for teaching multidigit arithmetic. Although many of the studies had interesting theoretical implications, for reasons of space we restrict our attention to the pre-post study of the methods class. Recall that the overall purpose of the measure was to provide feedback on teacher education, and in particular, for the specific methods class that was the intervention in this validation study. Our assumption was that the measure would be sensitive to changes in preservice teachers' mathematical proficiency for teaching multidigit arithmetic, which we further assumed would be influenced by the methods course.

The results of this study suggest that during the methods class preservice teachers developed knowledge and teaching self-efficacy beliefs, but they did not develop motivation beliefs. This finding is theoretically interesting because it provides direct evidence that an intervention can influence some but not all aspects of mathematical proficiency for teaching. Practically, it suggests that teacher education should look at a broader set of outcome measures than knowledge alone. Past research has shown that preservice mathematics students have increased preservice teachers' knowledge as measured by similar items (e.g., Copur-Gencturk & Lubienski, 2013), and moreover, that a mathematics content course for teachers increased motivation for fractions (Newton, 2009). However, no study to date has examined multiple constructs of mathematical proficiency for teaching with a shared mathematical focus, so it is plausible that because the measures are correlated, any teacher intervention that has an impact on knowledge would also have an impact on other facets of mathematical proficiency for teaching. Indeed, Jacobson (2017a) used a national sample of teacher education programs and showed that the length and timing of instruction-focused field experience has a similar pattern of effects across preservice teachers' mathematical content knowledge, as well as their beliefs about the nature of mathematics and mathematics learning. The pre-post validation study is the first study of which we are aware in which some aspects of mathematical proficiency changed in light of an intervention whereas others did not.

The study also drew our attention to the measure of value, which was already relatively high on the pretest. Unlike the measures for anxiety and self-concept of ability, there was little room on the scale to reflect an increase in preservice teachers' value for the mathematical knowledge reflected by the items. This finding has some practical implications for the measure of the value construct, namely, that a more extreme version of the questions should be asked in future surveys. In a follow-up study, we have piloted the revised phrasing, "This question is about one of the most important things elementary teachers need to know for teaching mathematics." The mean of the value scores on the new items is still on the upper half of the scale but much closer to the middle than were scores from the original wording. In addition to providing an empirical basis for improving the measure, the finding of a high mean value score has theoretical value in that it suggests the preservice teachers saw the knowledge items on the instrument as having relevance to their professional aspirations. This is important because the items are designed to represent realistic scenarios and tasks like those that teachers regularly face in the course of day-to-day instruction. In the next section, we examine validation efforts related to a different measure, which did not have the expected results.

Case 2: Measuring Children's Coordination of Length and Position on a Number Line

The second case we discuss deals with a different population and a different mathematical topic than the first case. However, the validation process has some parallels. We started with theory and with published assessments to write individual items. These were revised iteratively in response to the theoretical framework. Then, a series of think-aloud item response interviews were conducted to determine whether inferences drawn from responses to the questions accorded with inferences from what students said and did during interviews. We begin with a discussion of the theoretical construct and goals of the measure development project, then discuss the item response interviews. The results of the study revealed that the items were not in fact measuring what they had been designed to measure. However, the results also led us to consider the theoretical construct we aimed to measure in a new way, and provided new questions that have guided subsequent work.

Theoretical Construct and Measure Design

One major goal of mathematics instruction in the elementary grades is to develop students' reasoning about quantity so that they become "numerically powerful and proficient" (Chapin & Johnson, 2006, p. 1). There is an extensive amount of research and practitioner literature advocating students' use of representations of quantity—including number lines—during classroom instruction. Number lines have become prevalent in elementary classrooms based on an assumption that they are a good tool for building conceptual understanding (Teppo & van den Heuvel-Panhuizen, 2014). In spite of this apparent consensus, there is surprisingly little research on the ways in which students understand these common representations of quantity; that is, how students' underlying conceptions of quantity influence their use of quantitative representations. This gap in research is concerning because students do not seem to understand the representation well. For example, we consider rulers to be one form of number lines which are designed specifically for the purpose of measuring length. On a ruler task administered in the National Assessment of Educational Progress, only 20% of fourth-graders answered correctly (NAEP, n.d.). Motivated by these concerns, we sought to design a written instrument which would allow us to gather information about fifth-grade students' conceptions of number lines. Once designed, we intended to administer the instrument to a large number of students, and then perform psychometric analysis to study its validity as a tool to assess our construct of interest.

We hoped that our written instrument would allow us to measure students' thinking related to two possible conceptions of quantities on number lines—quantity as positions or as lengths. On any given number line, a number has one and only one position where it can be located. For example, any number placed on the number line will have only one correct position in relation to the given positions of 0 and 1. However, any number can also be shown as a line segment extending from 0 to that number. It is also important to understand that numbers interpreted as segments can shift; the segment of a specific length can be shown anywhere on the number line—including very far from its original position. We hypothesized that being able to "see" positions, understand that any position has an associated length between it and 0, and knowing that lengths can shift are all critical elements in unlocking the full potential of number line representations.

The work of refining our construct was particularly influenced by the Learning Mathematics through Representations (LMR) project, which developed a "research-based curriculum unit for the teaching and learning of integers and fractions in the elementary grades, using the number line as the principal representational context" (Learning Mathematics through Representations, n.d.). Prior to designing the curriculum, Saxe, Shaughnessy, Gearhart, and Haldar (2013) investigated fifth-grade students' ability to coordinate numeric and linear units on number lines. Their participants did not have a robust understanding of this coordination, although work with a distance context seemed to support such understanding (Saxe et al., 2013). Earnest (2015) administered an assessment which included number line tasks adapted from the LMR project to fifth- and eighth-grade students, then did a follow-up interview with a subset of those students. In the interviews, he found that of 23 students administered a particular number line task, only 6 successfully coordinated linear and numeric (positional) units (Earnest, 2015). Using these findings as an impetus, we sought to develop an instrument that would measure whether students "see" quantities on number lines as positions, as lengths, or as both. We were also interested in measuring whether students do or do not coordinate positional and lengthwise thinking when they work with number lines.

Over the course of three months, we developed a series of items that would elicit different types of thinking from students. Some items were designed to determine whether students could attend to the position of numerals on number lines. For example, students might be asked to mark where a specific number would be located on given number line (see Figure 3.3a). Other items were designed to determine whether students could solve problems which required them to attend to intervals on number lines (see Figure 3.3b). Further items were designed to determine whether students could coordinate the two types of thinking (position and interval) to solve problems (see Figure 3.3c). We hypothesized that students who thought of numbers only as positions or locations on number lines would be successful with the first category of problems, that students who could think of number as a length or interval would be successful with the second type of problem, and that students who could coordinate the two ways of thinking about numbers on lines would be able to solve the third type of problem.

As the items were being developed, we met weekly to discuss whether they would elicit the intended responses and what types of responses would indicate which type of thinking. We developed hypotheses for the various items—possible ways students would respond that would indicate different types of understanding. We also consulted an expert (a math education researcher who studies students' thinking about quantity) regarding the items and sequence, which led to further refinement. Figure 3.3 shows how one item, written to elicit student thinking about interval, evolved as a result of our discussions and feedback from the expert researcher. Initially, the item just asked students to show the number 4 on a line. It was then modified to include the word "length" and the numbers were changed so that none of the whole numbers encompassed by the line contained 4 as a digit. The item was then changed so that an interval was shown above the line and students would be asked to determine the length of that interval. The item was then changed so that the interval was shown as a rectangle on the line, and the numbers were changed to have 0 as a starting point. We also wrote scaled-up and scaled-down versions of the final item. For example, one scaled-up version did not have 0 as a starting point, and scaled-down versions included additional tick marks or had intervals of one unit.

Assumptions, Inferences, and Validation Study

The purpose of the measure was to accurately assess fifth-grade students' conceptions of number lines. We only considered elemental validity because the results from the first validation study disconfirmed our inferences. The elemental assumption was that the items reflect elementary students' conceptions of the number line in terms of a position meaning for numbers on the line, an interval meaning for numbers on the line, and the coordination of these meanings. One testable inference based on this assumption was that fifth-grade students' answers and reasons for their

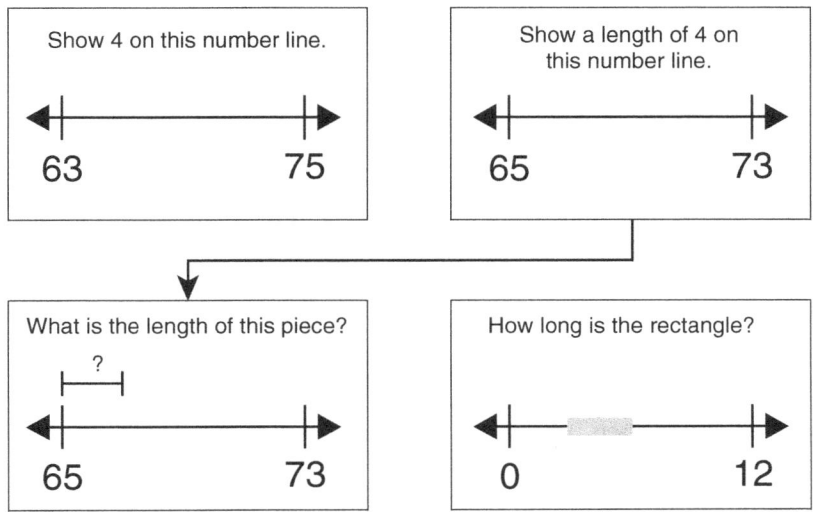

Figure 3.3 Evolution of an item.

answers would align. Those who answered position, interval, or coordination questions as we intended would be able to provide reasons that showed they had position, interval, or coordinated meanings for numbers on number lines. We sought to determine whether our questions would be understood by the students, whether they elicited the type of thinking we intended, and whether we could make reasonable conclusions about student thinking based on the types of answers they provided to their questions.

We conducted think-aloud item response interviews in order to test the inference. In the interviews, we asked students to answer a sequence of questions describing how they answered each question. We probed student thinking during the interviews to check whether we could make reasonable conclusions about thinking based only on written responses. We classified responses and were especially alert for false positives—responses that would lead us to conclude students have a certain understanding when they actually do not have that understanding—and false negatives—conclusions that students do not understand a concept when, in fact, they do. By testing the interpretive argument at the elemental level, these interviews were intended to provide validity evidence for the final measure.

Item Response Interviews

Participants of the interviews were 11 fifth-grade students from a rural school in the Midwest. Students were selected through convenience sampling (Creswell, 2012). On the days during which the study took place, the classroom teacher selected which students would participate in the interviews, based on her perception of students' ongoing success in math class. In other words, the teacher, not the researcher, decided which students would participate, and she selected students who she perceived to be doing well in mathematics to the extent that missing a portion of that day's instruction would not negatively impact their learning. As a result, the students who participated in the study do not represent the wide range of potential students in a typical fifth-grade math class. However, as our goal was critical examination of the items, not of students' conceptions, we did not perceive convenience sampling to be problematic in our study design.

Items were given to students one at a time, with each item printed on a separate page. Students were not given all of the items. During the sequencing of items, different question "paths" were determined, allowing for us to obtain data on all of the items without asking every student all of the items. Many of the items were scaled-up or scaled-down versions of previous items. If a student struggled with an item, they were given a scaled-down version. If they were successful with an item, they were sometimes given a scaled-up version so that we could better determine the depth of their understanding. Scaled-down versions often included extra tick marks on the line or smaller composite units. Scaled-up versions often had larger composite units or delineated segments of non-whole–number lengths. The interviewer (second author) also made in-the-moment decisions regarding which items to prioritize, based on student responses. Each interview lasted approximately 30 minutes. Students' work was video-recorded for later analysis of students' solution processes.

Early in the interview process, it became evident that our original plan of quickly identifying false negatives/positives, modifying the items into a written assessment, and then administering to a large sample was not reasonable. Students' understanding about numbers lines was much less explicit than we expected. Determining whether students were thinking about position or interval or coordinating the two was also difficult. Often, students answered in a way that we had anticipated, but when the interviewer asked follow-up questions, it became clear that their answer did not necessarily indicate the type of understanding we expected to accompany that answer. Further, we found many of the students to be very preoccupied with procedures they had been taught in their math classes, which made it difficult to ascertain their own, individual understandings. After conducting the interviews, we determined that it would be premature to administer a written assessment to a large sample. Instead, our focus turned to qualitative, in-depth analysis of the interviews, in the hopes of better understanding these 11 students' thinking about number lines. To provide support for this decision, we present a brief analysis of one student's responses to some of the items and a discussion of the complexity involved in determining how his answers did or did not fit within our hypothesized construct.

Sample Interview Findings

The work of one student, August (a pseudonym), is shown in Figure 3.4. When presented with the first item, August quickly stated that the rectangle was one unit long. To explain, he drew tick marks and labeled them with 27, 28, and 29. When asked where he saw a one-unit interval, August made a sweeping motion across the rectangle. We interpreted this as an indicator that August was attending to interval.

Given the second item, August drew tick marks across the line from left to right and said the rectangle was four units. Asked to explain, August responded that the rectangle would be one-third of the whole number line and again used a sweeping motion across the rectangle when he was referring to it. Asked where and how he saw four units, August first counted the three spaces, then restarted, counting the four tick marks from the beginning to the end of the rectangle. He then said that the rectangle would be a fourth of the whole line, once again using a sweeping motion.

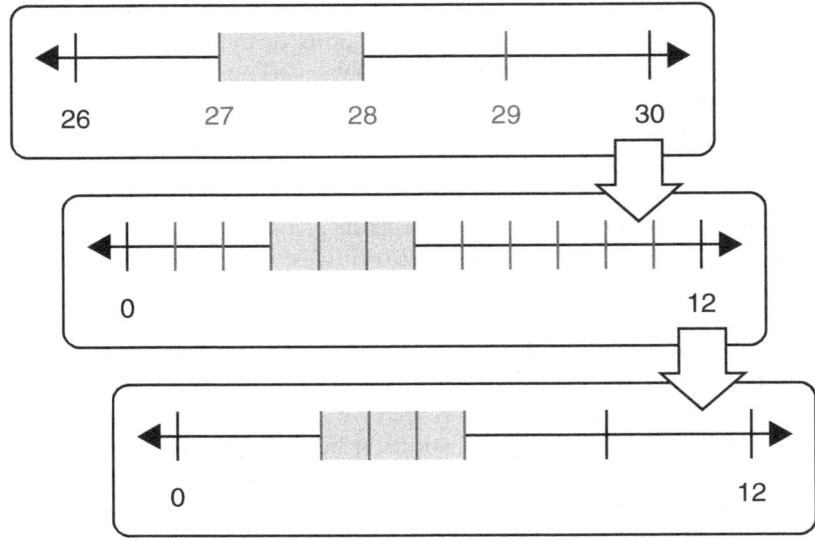

Figure 3.4 August's work in response to the question, "How long is the rectangle?"

He then stated, "There's just three different rectangles inside of one whole rectangle." Here August was coordinating the length of the rectangle with the length of the whole line, using his knowledge of fractions and multiplication/division to define the relationship between the two lengths. Yet he was still confused about what to count—tick marks or spaces—in order to determine the length of the rectangle, and he was still unsure whether the rectangle was one-third or one-fourth of the given line.

Noticing August's difficulty deciding, the interviewer presented him with a scaled-down version of the item, which had an additional tick mark between the rectangle and the 12. August quickly determined the rectangle would be one-fourth of the whole line, and then partitioned the rectangle into three parts. August said the rectangle would be three units long, "Because there's four [big] rectangles and . . . three [small] inside one. And then . . . four times three is twelve. That's how you get all the way up to twelve." Asked to clarify where he saw three, August responded, "Three different rectangles inside one big rectangle."

The spontaneous probing questions of the interviewer—in addition to the targeted use of a scaled-down item—allowed us to see that August was, indeed, attending to interval on these items, though he initially struggled to determine an answer. Further, he was able to coordinate the length of the rectangle with the length of the whole line, using multiplication and his knowledge of fractions to define the relationship between the two lengths. On a written assessment, however, it is likely we would have drawn drastically different conclusions about August's understanding. If he had answered "four" for the length of the rectangle, we may have concluded that he was not attending to length. If he had answered "three" we would not have known that particular ways in which he arrived at that answer. We likely would have had little (if any) knowledge about how August coordinated the length of the rectangle and the line and used computational reasoning to define and clarify the relationship.

In our interview data across the 11 students, results like the ones above were very common. The interview process allowed for probing questions that elicited further student explanations, which led to stronger conclusions about student understanding. At this point in our study, we concluded that we did not yet know enough about the various ways in which students make sense of number lines and the different ways in which those understandings manifest in written work.

Theoretical Implications of the Item Response Interviews

Our construct for understanding students' quantitative reasoning as applied to number lines was built on three categories—whether students attended to position, whether they attended to length, and whether they could coordinate position and length. In our analysis of August's work, however, we determined that looking at his responses in terms of these three categories was not particularly helpful in understanding his thinking. August's notions of position and length were intertwined. His attention to tick marks (and whether or which to count) showed he was attending to position, while his frequent sweeping motions across the number line showed he was also attending to length. He used both positions (tick marks) and lengths as he worked to determine the length of the given rectangle. Yet August was doing more than just coordinating position and length. He was also coordinating lengths with each other, considering how the length of the rectangle fit into the length of the larger line segment.

As we struggled to make sense of August's reasoning, we returned to the literature. This time we read more broadly. We found research in the area of measurement, particularly regarding how students unitize and iterate units (Lehrer, 2003), to be relevant to August's work, especially as we tried to determine what August was seeing as the unit. When August used the rectangle to measure how many times it would fit into the whole line then coordinated that with the number of pieces inside the rectangle, he appeared to be coordinating two levels of units. Our questions were not designed to elicit whether August had interiorized two or three levels of units (Hackenberg & Tillema, 2009); if they had been, we would likely be able to make more legitimate conclusions about what August's work said about how he was thinking about quantities on the number line. Further, we could have incorporated questions into our interview that would have allowed us to analyze whether (and when) August was thinking additively and/or multiplicatively and whether he was able to see units as being simultaneously embedded within other units (Ulrich, 2015, 2016). Finally, we began to consider that length and position on number lines may not be mutually exclusive categories—especially from the perspective of elementary students—and we are working to connect our ideas about these categories to literature about how students conceptualize discrete and continuous quantities (Clement, 1989; Johnson, 2012). We are currently working on developing a theoretical framework that incorporates all of these ideas. Our construct has become much more complex, and we are conducting further qualitative studies on this topic.

In pursuit of an instrument that would allow us to measure students' conceptions of quantity, as exhibited on number lines, we discovered that this construct was much more complex than we anticipated. The difficulty in analyzing August's (and others') work indicated that our construct, as theorized, was not sufficient. One perspective might frame our work as being unsuccessful—we sought to develop an instrument that could be used to assess students' number line concepts and we did not succeed in doing so. However, engaging in this work opened the door for further development of our theoretical framework, which is informing our future work. Doing this work has allowed us to better understand the literature in the field as well as student data from future projects in a way that we would not have been capable of had we not engaged in this study.

Discussion and Conclusions

In this chapter, we have discussed two cases of measure validation. Our thesis is straightforward—instrument validation has underused potential as a research method for the field of mathematics education. We presented two cases where measure validation studies had significant theoretical implications leading to new research questions and thereby advancing the field. We showed these theoretical benefits both in the first case, where the validation results were generally positive, and in the second case, where the results were negative.

We infer that the potential contribution of validation studies is not widely recognized because so few validation studies have been published in mathematics education (see Bostic, Krupa, Carney, & Shih, 2019; Bostic, Lesseig, Sherman, & Boston, in press). Part of this may be due to the perception that validation of measures is 'development' work and not research. Yet, when validation studies are framed within theory then results of these studies can speak directly to theoretical concerns. We have found that the framework for measure validation proposed by Schilling

(2004, 2007) has utility in this regard. All current models of measure validation share an argument-based approach, but Schilling's approach strikes a middle ground between foregrounding the argument (as Kane's approach to validation does) and foregrounding the validity evidence (as the *Standards* approach does). The key difference, in our view, is that Schilling's approach explicitly requires testable inferences, which is just another name for a falsifiable hypothesis. Falsification is a critical component of validation work. By making falsification the focus, the theoretical implications of validation studies are clear, no matter whether the results are positive or negative, as the two cases we examined in this chapter demonstrate.

What then are the implications for the field? We believe that the generality of the phenomena of interest in mathematics education is understudied. Measures are an excellent means of investigating generality, both across items within the measure and across different individuals within the population. Measures require test developers to operationalize their inferences in every item, and this requires stating hypotheses in very specific ways. Once a set of items has been developed, measures allow researchers to test generality of the phenomena across different groups of people (or across the same group at different points in time). By framing measure validation in terms of theory, validation studies can advance theory by examining the generality of phenomena of theoretical interest. Mathematics education as a field can only benefit from more measure validation work undertaken by mathematics education researchers. Such efforts have great potential to help researchers, policymakers, and other stakeholders outside of mathematics education understand and value the phenomena that our field values. An increased focus on measure validation can improve the rigor of our field by addressing questions of generality directly, and increase our field's influence on education in society.

The burden to change the culture of mathematics education falls on authors, reviewers, and editors alike. Authors who conduct validation research should frame their results in ways that clearly contribute to the field, for example, by detailing how theory is advanced. These reports should be submitted to mathematics education journals with a thoughtful cover letter asking editors to send such papers out for review in light of the potential contribution. Reviewers who encounter such work should provide feedback to authors that helps identify the aspects of the validation research that can contribute to the field more broadly than the specific measures in question. And rather than using measure development as a heuristic for desk rejection, editors of mathematics education research journals must think more inclusively about measure development as a research methodology. In the absence of such changes, the time is ripe for a new journal focused on measurement in mathematics education.

References

American Educational Research Association, American Psychological Association, & National Council on Measurement in Education. (2014). *Standards for educational and psychological testing.* Washington, DC: American Educational Research Association.

Baumert, J., Kunter, M., Blum, W., Brunner, M., Voss, T., Jordan, A., . . . Tsai, Y. (2010). Teachers' mathematical knowledge, cognitive activation in the classroom, and student progress. *American Educational Research Journal, 47*(1), 133–180.

Bostic, J., Krupa, E., Carney, M., & Shih, J. (2019). Reflecting on the past and thinking ahead in the measurement of students' outcomes. In J. Bostic, E. Krupa, & J. Shih (Eds.), *Quantitative measures of mathematical knowledge: Researching instruments and perspectives.* New York, NY: Routledge.

Bostic, J., Lesseig, K., Sherman, M., & Boston, M. (in press). Classroom observation and mathematics education research. *Journal of Mathematics Teacher Education.*

Cai, J. (2005). US and Chinese teachers' constructing, knowing, and evaluating representations to teach mathematics. *Mathematical Thinking and Learning, 7*(2), 135–169.

Chapin, S. H., & Johnson, A. (2006). *Math matters: Understanding the math you teach.* Sausalito, CA: Math Solutions.

Clement, J. (1989). The concept of variation and misconceptions in Cartesian graphing. *Focus on Learning Problems in Mathematics, 11*(1–2), 77–87.

Copur-Gencturk, Y., & Lubienski, S. T. (2013). Measuring mathematical knowledge for teaching: A longitudinal study using two measures. *Journal of Mathematics Teacher Education, 16*(3), 211–236.

Creswell, J. W. (2012). *Educational research: Planning, conducting, and evaluating quantitative and qualitative research.* Upper Saddle River, NJ: Pearson Education, Inc.

Earnest, D. (2015). From number lines to graphs in the coordinate plane: Investigating problem solving across mathematical representations. *Cognition and Instruction, 33*(1), 46–87.

Hackenberg, A. J., & Tillema, E. S. (2009). Students' whole number multiplicative concepts: A critical constructive resource for fraction composition schemes. *Journal of Mathematical Behavior, 28,* 1–18.

Hill, H. C., Rowan, B., & Ball, D. L. (2005). Effects of teachers' mathematical knowledge for teaching on student achievement. *American Educational Research Journal, 42*(2), 371–406.

Jacobson, E. (2013). *Mathematics teachers' professional experience and the development of mathematical proficiency for teaching* (Unpublished doctoral dissertation). Athens, GA: University of Georgia.

Jacobson, E. (2017a). Field experience and prospective teachers' mathematical knowledge and beliefs. *Journal for Research in Mathematics Education, 48*(2), 148–190.

Jacobson, E. (2017b). *Sample items from a survey measure of mathematical proficiency for teaching.* [Figure]. https://doi.org/10.6084/m9.figshare.4793437.v1

Jacobson, E., Aydeniz, F.,* Creager, M.,* Diaga, M.*, & Uzan, E.* (2018). Mathematics teachers' knowledge and productive disposition for teaching: A framework and measure. In G. Stylianides & K. Hino (Eds.), *Research advances in* *the mathematical education of preservice elementary teachers* (pp. 187–203). New York, NY: Springer.

Jacobson, E., & Izsák, A. (2015). Knowledge and motivation as mediators in mathematics teaching practice. *Journal of Mathematics Teacher Education, 18*(5), 467–488.

Jacobson, E., & Kilpatrick, J. (2015). Understanding teacher affect, knowledge, and instruction over time: An agenda for research on productive disposition for teaching mathematics. *Journal of Mathematics Teacher Education, 18*(5), 401–406.

Jacobson, E., & Svetina, D. (2019). Prescribing structure for validation arguments: Elemental, structural, and ecological validity. *Applied Measurement in Education, 32*(1), 43–59. doi:10.1080/08957347.2018.1544137

Johnson, H. (2012). Reasoning about variation in the intensity of change in covarying quantities involved in rate of change. *The Journal of Mathematical Behavior, 31*, 313–330.

Kane, M. T. (2004). Certification testing as an illustration of argument-based validation. *Measurement, 2*(3), 135–170.

Kane, M. T. (2013). Validating the interpretations and uses of test scores. *Journal of Educational Measurement, 50*(1), 1–73.

Learning Mathematics for Teaching Project. (n.d.). *Content and tasks measured.* Retrieved from http://sitemaker.umich.edu/lmt/content

Learning Mathematics through Representations. (n.d.). *Overview.* Retrieved from http://culturecognition.com/lmr/

Lehrer, R. (2003). Developing understanding of measurement. In J. Kilpatrick, W. G. Martin, & D. Schifter (Eds.), *Research companion to the principles and standards for school mathematics* (pp. 179–192). Reston, VA: National Council of Teachers of Mathematics.

National Center for Education Statistics. (n.d.). *NAEP questions tool, item 2003-4M6 #18.* Retrieved from https://nces.ed.gov/NationsReportCard/nqt/Search

Newton, K. J. (2008). An extensive analysis of preservice elementary teachers' knowledge of fractions. *American Educational Research Journal, 45*(4), 1080–1110.

Newton, K. J. (2009). Instructional practices related to prospective elementary school teachers' motivation for fractions. *Journal of Mathematics Teacher Education, 12*(2), 89–109.

Peterson, C. H., Peterson, N. A., & Powell, K. G. (2017). Cognitive interviewing for item development: Validity evidence based on content and response processes. *Measurement and Evaluation in Counseling and Development, 50*(4), 217–223.

Saxe, G. B., Shaughnessy, M. M., Gearhart, M., & Haldar, L. C. (2013). Coordinating numeric and linear units: Elementary students' strategies for locating whole numbers on the number line. *Mathematical Thinking and Learning, 15*(4), 235–258.

Schilling, S. G. (2004). Conceptualizing the validity argument: An alternative approach. *Measurement, 2*(3), 178–182.

Schilling, S. G. (2007). Generalizability and specificity of interpretive arguments: Observations inspired by the commentaries. *Measurement, 5*(2–3), 211–216.

Schmidt, S. H., Houang, R., & Cogan, L. S. (2011). Preparing future math teachers. *Science, 332*(6035), 1266–1267.

Shulman, L. S. (1986). Those who understand: Knowledge growth in teaching. *Educational Researcher, 15*(2), 4–14.

Silver, E. (2004). Ella Minnow Pea: An allegory of our times? *Journal for Research in Mathematics Education, 35*(3), 154–156.

Teppo, A., & van den Heuvel-Panhuizen, M. (2014). Visual representations as objects of analysis: The number line as an example. *ZDM, 46*(1), 45–58.

Ulrich, C. (2015). Stages in constructing and coordinating units additively and multiplicatively, part 1. *For the Learning of Mathematics, 35*(3), 2–7.

Ulrich, C. (2016). Stages in constructing and coordinating units additively and multiplicatively, part 2. *For the Learning of Mathematics, 36*(1), 34–39.

Wigfield, A., & Eccles, J. S. (2000). Expectancy-value theory of achievement motivation. *Contemporary Educational Psychology, 25*(1), 68–81.

Wigfield, A., & Meece, J. L. (1988). Math anxiety in elementary and secondary school students. *Journal of Educational Psychology, 80*(2), 210.

4 Gathering Validity Evidence Using the BEAR Assessment System (BAS)

A Mathematics Assessment Perspective

Mark Wilson and Diane B. Wilmot[1]

Introduction

The audience for this chapter is mathematics educators and researchers (1) who want to better understand what good validity evidence for assessments looks like and/or (2) who want to develop validity evidence for their mathematics assessments. The current central conception of validity is that it should be embodied in an argument that assembles validity evidence to support the uses and interpretations of the outcomes from an assessment. As Kane (2006) stated:

> To validate an interpretation or use of measurements is to evaluate the rationale, or argument, for the proposed conclusions and decisions ... Ultimately, the need for validation derives from the scientific and social requirement that public claims and decisions be justified.
>
> (p. 17)

This position will be adopted, and expanded, in this chapter.

What is emphasized in this chapter is the importance of the establishment of the 'claim' of the argument, sometimes called the 'hypothesis' to be argued for. This may seem an odd thing to emphasize, as surely there can be no argument without a claim, so this must have been present all along. The contention of this chapter is that, first, the argument will be stronger and more convincing when the initial hypothesis is richer, with more structural aspects that can be compared to empirical evidence. Second, that the argument must be shaped by the nature of those structural aspects.

What is needed for the interpretive argument can be seen as being composed of four parts, as follows:

1. A framework that allows teachers and others to interpret student development in the variable under assessment;

2. Methods of gathering data that are acceptable and useful to appropriate audiences;
3. A way to value what we see in student work; and
4. A technique of interpreting data that allows meaningful reporting to multiple audiences.

As will be described below, an approach called the BEAR Assessment System (BAS) supports this structure. Note that we are emphasizing the context of a teacher interpreting instructional assessments, as we perceive that this is the focus of this volume, but indeed this framework can be applied broadly in education and across the social sciences, in the pursuit of measurements of many sorts, in achievement, attitudinal, and behavioral domains.

Traditionally, validity has been seen as composed of several different components—as described in the *Standards for Educational and Psychological Testing*, commonly referred to as the *Standards* (American Educational Research Association, American Psychological Association, & National Council on Measurement in Education [AERA, APA, & NCME], 2014).

Over time, the nature and number of these components has evolved: for example, in the 1985 edition they were described as evidence related to *criterion*, *content*, and *construct* aspects of validity; and in the 2014 edition they were noted as evidence based on the following strands of evidence: *test content*, *response processes*, *internal structure*, *relations to other variables*, *consequences of testing*, as well as the validity issue of *fairness*. In this chapter, we will use the latest (2014) version—these strands are described in a later section of the chapter, so we will not elaborate here. The focus in this chapter is on how these can be seen as important, and sometimes overlapping, parts of an argument supporting the validity of an instrument, and hence each of these categories will be used, in turn, as a source of possible evidence, either for or against the planned usage of the instrument.

In order to avoid empty theorizing, we will contextualize our arguments using a specific mathematics assessment project that we worked on previously: thus, in the first section below, we will describe this exemplary context—the Functions Learning Progression (FLP) project (Wilmot et al., 2011). In the second section, we will describe the general approach to test construction that will be used here (i.e., the BAS) and use the FLP project to illustrate how the BAS principles can be applied in a concrete context. The next section will describe how the validity evidence can be gathered and used following the BAS approach, again contextualized within the FLP project—this will be organized following the order of the list of the strands of validity evidence given above. Finally, the conclusion will summarize the work, and comment on future directions.

Brief Description of the Functions Learning Progression Project

A group of measurement, mathematics, science, and policy experts, funded by Berkeley Futures Project (BFP), met regularly during the 2006–2007 academic year to discuss the literature and research regarding to college readiness standards for students and policies. One aim of these gatherings was to plan for an assessment of college readiness. The group decided to focus the content of the assessment around what they saw as an important topic within the mathematics curriculum—making connections across multiple representations of mathematical functions. This was seen as being relevant to numerous college majors beyond the obvious mathematical ones (e.g., architecture design, computer programming, business, demographics, social policy, economics, and women's studies). The topics of functions and functional relationships are core algebraic concepts that span kindergarten through 12th grade (Chief Council of State School Officers [CCSSI], 2010; National Council of Teachers of Mathematics, 2000), and also are a significant focus of university mathematics. This latter point was based on recommendations from groups such as the California State University Academic Senate (1997), which recommended that new college students need to be familiar with representations of functions, such as using graphs, tables, variables, and words, because students in many different disciplines have to use and understand mathematical functions. In addition, there was a focus on functions because there is a relatively large knowledge base of student learning in this domain (a small sample: Leinhardt, Zaslavsky, & Stein, 1990; Moschkovich, Schoenfeld, & Arcavi, 1993; Piaget, Grize, Szeminska, & Bang, 1977). The body of research thus located was used to develop the learning progression.

The field-test version of the FLP test was administered to 2,356 students in 125 classrooms across the United States, and spanning grades 6 to 12. Some results reported below were based on smaller subsets of this total set, as several different forms were used in the field test, in order to capture data about more items than could reasonably be given to students is a single sitting (Wilmot et al., 2011). The research design and procedures were examined and approved under the UC Berkeley IRB process managed by the Berkeley Committee for the Protection of Human Subjects.

The BEAR Assessment System

The BEAR Assessment System is based on the idea that good assessment addresses these considerations through four "building blocks"—the *construct map*, the *items design*, the *outcome space*, and the *measurement model*. See Wilson (2004) for a detailed account of an instrument development process that works through these steps. Below we take up each of these building blocks in turn—these will be used in the next section to illustrate the gathering of validity evidence.

Building Block 1: The Construct Map

We take a developmental perspective of student learning, and this is embodied by the *construct map* (e.g., see Figure 4.1). Learning is conceptualized not simply as a matter of acquiring quantitatively *more* knowledge and skills, but as progress toward higher levels of competence as new knowledge is linked to existing knowledge and as deeper understandings are developed from and take the place of earlier understandings. To use the BAS in any given area it is assumed that learning can be described and mapped as progress in the direction of qualitatively richer knowledge, higher-order skills, and deeper understandings.

Variables are derived in part from research into the underlying cognitive structure of the domain and in part from professional opinion about what constitutes higher and lower levels of performance or competence, but are also informed by empirical research into how students respond to instruction or perform in practice (National Research Council [NRC], 2001). To more illustrate more clearly the nature of a progress variable, consider the Functions example introduced in the previous section.

The FLP was developed using the SOLO taxonomy (Biggs & Collis, 1982), a general assessment–development framework that is used to assess student ability to make connections within and across topic areas (Biggs, 1999). The SOLO taxonomy was readily applied to the area of mathematical functions, as the aim was to assess student ability to do just that, for the topic of multiple representations of mathematical functions.

The usual SOLO taxonomy has five levels of student understanding. The project adapted the taxonomy by adding a lower and initial level called *prealgebraic*. The FLP consists of a single construct map,[2] shown in Figure 4.1. In this figure, student responses, going from bottom to top, represent increasing sophistication of student understanding of the connections among mathematical functions. For details about these levels and expected student responses, see Wilmot et al. (2011, pp. 265–266).

Creating the construct map is not a trivial task. But having succeeded in adapting this approach to a given curriculum, the instructor will be well situated to address many of the issues raised in the introduction to this chapter. This approach assumes a match between instruction and assessment, which we address next.

Building Block 2: The Items Design

The match between the instruction and assessment in the BAS is established and maintained through two major parts of the system: the construct maps, described above, and assessment tasks or activities, described in this section. The plan for these is called the *items design*. The assumption

Functions Learning Progression: Making Connections Across Multiple Representations of Mathematical Functions		
Levels of Sophistication	Student is able to:	Response to items
Extended Abstract (5)	Make connections not only within a given subject area, but also beyond it, able to generalize and transfer the principles and ideas underlying the specific instance.	• Predict, explain and synthesize their understanding to a real-world context. • Solve non-routine problems including non-algorithmic functions.
Relational (4)	Demonstrate understanding of the significance of the parts in relation to the whole.	• Compare/contrast information given in multiple representations of functions to demonstrate understanding of the content. • Recognize which representations to choose for the context. • Select representations and move fluently among them to achieve a solution.
Multi-structural (3)	Make a number of connections, but the metaconnections between them are missed, as is their significance for the whole.	• Make connections across more than two representations (e.g., tabular, symbolic, graphic and verbal representations). • Recognize more than one feature of a functional relationship.
Unistructural (2)	Make simple and obvious connections, but their significance is not grasped.	• Make pairwise connections between representations.
Prestructural (1)	Acquire bits of unconnected information, which have no organization and make no sense.	• Interpret graphs where both variables have to be interpreted, or where time is independent variable. • Demonstrate understanding of a function in one representation
Prealgebraic (0)	Acquire prerequisite skills	• Demonstrate understanding of: • Functional dependence, where a change in one variable affects another variable. • Continuous variables (time, distance) and dichotomous variables (hot or cold). • Relative order and measurement to define variables.

Figure 4.1 The Functions construct map.

(Note: read from bottom to top.)

here is that the framework for the curriculum and instruction must be one and the same. This is not to imply that the needs of assessment must drive the curriculum, nor that the curriculum description will entirely determine the assessment, but rather that the two, assessment and instruction, must be aligned—they must both be designed to accomplish the same thing, the aims of learning, whatever those aims are determined to be.

In order to make this alignment concrete the assessment tasks need to reflect the range and styles of the instructional practices in the curriculum. They must have a place in the "rhythm" of the instruction, occurring at places where it makes instructional sense to include them, usually where instructors need to see how much progress their students have made on a specific topic.

Doing so brings the richness and vibrancy of curriculum development into assessment, and also brings the discipline and hard-headedness of assessment data into the design of instruction.

The variety of assessment tasks used by the BAS can range widely, including individual and group "challenges"; data interpretation questions; and tasks involving student reading, laboratory, or interactive exercises. The FLP test included 12 constructed response items that asked students to write their responses with words, algebraic formulas, and/or numbers. An example item, the Hexagon Pattern Task, is shown in Appendix 4.A.

Building Block 3: The Outcome Space

For information from the assessment tasks and the BAS analysis to be useful to instructors and students, it must be couched in terms that are directly related to the instructional goals behind the progress variables. Open-ended tasks, if used, must be designed so that they can be quickly, readily, and reliably judged and categorized according to the construct map. This judging of responses can be carried out by people, such as teachers, students themselves, readers, teaching assistants and instructors, or by machine, say, using web-based interfaces with real-time delivery of instructional material and feedback, or more traditional machine-readable answer sheets based on multiple choice items. The rules on which these judgments are based may be simple and ordered, such as is shown for the Hexagon Task, or they can be complex, involving complicated subparts of responses, etc.—either way, they must suffice to result in a level of the construct map.

When scoring guides are used, teachers and students need concrete examples—which we call "exemplars"—of the rating of student work, examples of what an instructor might expect from students at varying levels of development along each variable. They are also a resource to understand the rationale of the scoring guides. Actual samples of student work, scored and moderated by those who pilot-tested the BAS in the FLP test, are available for each activity. These illustrate typical responses for each score level, as well as atypical responses that exercise the raters' skills: an example for the Hexagon Task is shown in Figure 4.2.

Building Block 4: The Wright Map

Our approach on this technical end of measurement is to use the special set of statistical models in item response theory (IRT), known as Rasch

Connections Construct	GP-1 Exemplars
EA - Extended Abstract. Applies knowledge of functions for hexagons to a generalizable formula for n-gons	Symbolic function for n-gon (1g) is correct and related to the symbolic function for hexagon (1c)
R - Relational - Relates information from hexagons to determine functions for n-gons	Verbal function for n-gon (1e/1g) is correct and related to verbal function for hexagon (1b)
MS - Multistructural - More than 2 functions (tabular, verbal, symbolic) are clearly connected	Table (1a), verbal description (1b) and equation (1c) are correct and connected
US - Unistructural - 2 functions are connected	Table (1a) and equation (1c) are connected OR Table (1a) and verbal description of 100 hexagons are connected (1b) (Note: Verbal description must demonstrate knowledge beyond "adding four")
PS - Prestructural - 1 function correct	Table (1a) is correct.
PA - Prealgebraic- 1 function partially correct	Table (1a) is partially correct (Note: Student may mention "adding four" as the only substance to the problem)
OT - Off Target	Nothing correct. Makes no sense.
DK - Don't Know	Writes "I don't know"
NR- No Response	Blank response

Figure 4.2 Scoring guide for the Hexagon Pattern Task.
(Note: read from bottom to top.)

models (Rasch, 1960/1980), and as described by Adams, Wilson, and Wang (1997) and Adams, Wilson, and Wu (1997). These are examples of measurement models well-developed enough for use in classroom-based assessment in a fairly routine and feasible way. The output from these models can be used as quality-control information, and to determine where individual students fall on a construct map such as FLP's Functions variable. Results from this approach were used to validate and calibrate the Functions progress variable. The first step is to create an empirical version of the construct map, such as the one shown in Figure 4.1: This new version is called a *Wright map*, and this is illustrated in Figure 4.3.

Looking at Figure 4.3, the first column (from the left) shows the levels on the construct map. The second column indicates the estimated ability

College Readiness Assessment – Wright Map

Level on the Connections Construct Map	Estimate	Distribution of student proficiency (n = 688)	Distribution of item difficulty (n = 8)							
			HEXAGON PATTERN	EQUIVALENT FUNCTIONS	EDUCATIONAL ACCOMPLISH-MENTS	GENDER GAP	COST OF POSTAGE	STAIRCASE TOOTHPICK	24-HOUR CRUDE OIL	100 YEARS CRUDE OIL
Extended Abstract (EA) / Relational (R)	4	X								
	3	XXX	EA		EA				RA	EA
	2	X XXX	R	R	R	MS	R	MS US	R	R
Multistructural (MS) / Unistructural (US)	1	XXXXXX XXXXXXXX XXXXXXXXXXXXXXX XXXXXXXXXXXXX -XXXXXXXXXXXXXXXX XXXXXXXXXXXXX XXXXXXXXX XXXXXXXXX	MS US		MS	US	MS		MS	MS
	0			MS US	US	PS	US PS	PS PA	US	US
Pre-structural (PS) / Pre-algebraic (PA)	−1	XXXXX XX XXX	PS PA	PA PS	PA	PA	PA		PS PA	
	−2	X X X								PS PA

EACH X REPRESENTS 4 STUDENTS, EACH ROW REPRESENTS .255 LOGITS

Figure 4.3 Wright map for the FLP Functions construct.

of the students in logit units. The third column on this map shows the measured distribution of a set of students who responded to the Functions items in field trials of the test. Then there are eight more columns, one for each of the eight items (named at the top of the eight columns). The calibrated difficulty of the thresholds between the levels of each of the tasks is shown within each column. Item response modeling can be used to locate a student or describe an entire class, as well as generate fit statistics and other indices for how well levels specified by the model fit the classroom data. In Figure 4.3, for example, note that the top levels for the "24-Hour Crude Oil" and the "100-Years Crude Oil" tasks (Note: these are denoted by "EA"—see top left of Figure 4.2 for the meaning of this abbreviation, and the rest as well) are above the highest student location—this means that it is quite unusual for students to respond at those levels. Looking down at the bottom of the Wright map, it can be seen that it is highly likely that some students responded at level PA or below. Most students have responded at about the US or MS levels. At the far left of the Wright map, one finds the three levels of the construct map that were eventually seen as being distinguishable by the project participants (more on this below).

The formal nature of the statistical models that were used here, and their flexibility, allows technical challenges inherent in the classroom assessment situation to be addressed, such as the maintenance of instructor rating consistency and the maintenance of a meaningful scale throughout the school year. This puts richer information into the hands of instructors in the classroom. Advice can be developed by and for teachers, and many types of maps are available, derived from analysis of student data collected in coursework.[3]

Wright maps can be used to record and track student progress and to illustrate the skills a student has mastered and those that the student is working on. By placing students' performance on the continuum defined by the map, teachers can illustrate students' progress with respect to the goals and expectations of the course. The maps, therefore, are one tool to provide feedback on how students as a whole are progressing in the course. They can also be used to inform instructional planning. Wilmot (2008) interviewed a small sample of teachers to examine the usages that they made of these the Wright maps and associated reports of results, and found that they facilitated the following activities: (1) checking student progress, (2) making instructional decisions, (3) seeing learning as a trajectory, and (4) planning curriculum across years. Details and examples are contained in Chapter 5 of her dissertation (Wilmot, 2008, pp. 126–131).

The Design Validity Argument: Using the BAS Approach

To illustrate how the BAS approach will function for the purpose of gathering validity evidence, we will follow the strands of validity evidence, as included in the *Standards* (AERA et al., 2014) mentioned above.

A summary of how the four building blocks of the BAS relate to each of the strands is shown in the design validity evidence matrix shown in Figure 4.4. As you can see, not every building block relates to each evidence strand, but each relates to several: For each evidence strand, more than one building block can be relevant.

In what follows, the FLP example is used to illustrate cells in the matrix, but, as will usually be the case, not everything indicated in the matrix will be relevant to this particular example—when that is the case, we make some general remarks about the cells that do not show up in the example. Note that in the sections below we are using the term "measurer" to denote whoever is looking into that particular form of validity evidence. Most often it will be an assessment developer, but sometimes it might be an assessment user, so we have kept to the more general term.

Evidence Based on Instrument Content

To compile evidence based on an instrument's content, the measurer must engage in "an analysis of the relationship between a test's [instrument's] content and the construct it is intended to measure" (AERA et al., 2014), and interpret that analysis in an argument concerning the validity of using the instrument. One way to see this is to think of the "content" of the instrument as describing the construct and instrumentation *hypotheses* that the instrument embodies. Developing such a set of hypotheses is exactly what has been described above in the section labeled "The BEAR Assessment System." In particular, the Functions construct of the FLP project has been illustrated through a construct map in Figure 4.2, an example item has been provided in Appendix 4.A, and a scoring guide has been shown in Figure 4.3. If the measurer carries out the steps outlined there (and described in much more detail in Wilson (2004, chapter 2)), then the measurer will have in place exactly such a set of content hypotheses. Much of what follows in the next several subsections amounts to an investigation of possible evidence supporting or contradicting those hypotheses.

Considering the FLP project, and looking at the second row[4] of the evidence matrix (Figure 4.4), consider the second cell[5] of that row:

Documentation of development using four building blocks.

Indeed, the roles of the four building blocks have been described as part of the development process (Wilmot et al., 2011, pp. 264–271), and the content of the items and scoring guides do manifestly match the construct map design. Consider now the third and fourth[6] cells of that same row:

Item set has the construct features that were planned—validated by: inspection, or panel of judges.
Outcome Space, and scoring guides, have the construct features that were planned—validated by: inspection, or panel of judges.

Strands of Validity Evidence	Constructs	Items Design	Outcome Space	Measurement Model
Content: Construct maps, items Design, Outcome Space	Documentation of development using four building block.	Item set has the construct features that were planned—validated by; inspection, or panel of judges.	Outcome space, and scoring guides, have the construct features that were planned—validated by: inspection, or panel of judges.	Chosen measurement model is consistent with construct maps, etc.
Response processes: Think-alouds, exit interviews	Overall, the construct levels are found to all be present and distinct: Can "hear" different levels from students.	*Each item (or item-type)...* prompts information that is consistent with the construct.	*Each item (or item-type)...* prompts information that is consistent with the construct levels (though specific items may generate only a subset).	Examine student statements for consistency or inconsistency with the assumptions of the measurement model.
Internal structure: Test level	On the Wright map, items representing the construct levels fall into reasonably different and ordered bands, as expected.	On the Wright map, items are dispersed in a reasonable way.	On the Wright map, item thresholds are dispersed in a reasonable way.	Test fits model reasonably well: Examine item fit statistics. Person reliability is within reasonable levels.
Internal structure: Item level		*Each item (or item-type)...* falls within the expected band for its appropriate level on the Wright map.	*For each item, each category threshold)...falls within the* expected band for its appropriate level on the Wright map.	Items fit construct map reasonably well: Examine category location table for *each item*, etc.
Internal structure: Person level	On the Wright map, students are dispersed as expected.	On the Wright map, students are reasonably well-matched to items.	On the Wright map, students are reasonably well-matched to item thresholds.	Students fit the model reasonably well: Examine student fit statistics.
Relations to other variables	The relationship between the estimates and the other variable is as expected.	The relationship between the item scores and the other variable is as expected.		
Consequences	Training on the construct helps teachers teach the topic, and students learn it. Negative consequences are not found.	Training using the items helps teachers teach the topic, and students learn it. Negative consequences are not found.	Training using the outcome space and the scoring guides helps teachers teach the topic, and students learn it. Negative consequences are not found.	Reporting based on the construct helps teachers teach the topic, and students learn it. Negative consequences are not found.
Fairness	Test operates in the same way for different groups: Examined by study of differential test functioning.			Items (or item types) operate in the same way for different groups: Examined by study of differential item functioning.

Figure 4.4 The design validity evidence matrix.

We note that the products of that process (i.e., both items and the outcome space) were not subjected to a review by an independent set of judges, but, as noted above, an advisory group of experts did oversee the whole process, including critiquing the construct maps, items, and scoring guides. Consider the fifth cell of that same row: "Chosen measurement model is consistent with construct maps, etc." As noted above, the Rasch measurement model was used for analyses, and this does indeed match the intent of the designers of the constructs (the particular form of the Rasch model that was used is suitable for polytomous [i.e., nondichotomous] data).

Armed with a sound construct map, it provides the fundamental premise about content, and with that in hand, the path to developing strong evidence has a strong first step.

Evidence Based on Response Processes

To compile evidence based on response processes, the measurer should engage in a detailed analysis of individual responses in so-called "cognitive lab" interview exercises,[7] either while the individuals are taking the instrument, a "think-aloud," or just after, in an "exit interview" (Nisbett & Wilson, 1977). The evidence thus gathered, as well as evidence gathered in separate studies after the instrument has been developed, can all be used as part of the validity evidence for the instrument. It is important to appreciate that this form of validity evidence can be used both formatively and summatively, and that both forms are relevant validity evidence.

Like the comments for the evidence related to content, the evidence gathered under this heading is extremely important for the soundness of the instrument: It sheds light on many of the expectations that are to be compared to outcomes during the investigations into other aspects of validity. This evidence can be used formatively to improve the instrument—in this case, documenting the improvements that have been made, along with their evidentiary justifications, contributes to evidence for the validity of the development process. In addition, for final-form items, the response process information provides direct validity evidence for the item set.

Considering the FLP project, and looking at the third row of the evidence matrix, Wilmot and her colleagues (2011) describe a thorough process of investigation of student responses to the items, and document a set of student responses to the Hexagon Pattern Task that show clearly that students do indeed make comments that are related to the levels of the construct map. They do not document the many steps taken to improve the items and the scoring guides using the think-aloud information, though a summary is available in Wilmot (2008). No evidence regarding the suitability of the measurement model from the response process information was reported.

Having response process evidence ties the validity of the outcomes from the instrument closely to the students and their experiences. Focusing on the evidence related to the construct map will help with assembling validity evidence, but the response process evidence can inform much more than that.

Evidence Based on Internal Structure

To compile evidence based on internal structure, the measurer must first establish what is the *intention* for the internal structure. Although this idea of intended structure may not be generally explicitly discussed for instruments designed in many areas, if the measurer has followed the steps of the BAS outlined above, then indeed there must be an internal structure that is expected—it is the structure of the construct maps. If the measurer builds this construct following the building blocks in the BAS there will be an expectation that the levels of the construct will be distinguishable, and will be in a certain order, running from high to low, easy to difficult, or negative to positive, as the case may be. The evidence that the content expectations have been fulfilled will reside in the analysis of a field test of the data, analyzed using the measurement model, and displayed in a Wright map. The concordance between the theoretical expectations in the construct map and the empirical results in the Wright map reflects the support that the evidence offers.

As there are several aspects of internal structure present in a model as complex as the Rasch model, the evidence can be examined at several levels: (1) test level, which considers how well the empirical evidence matches the intention for the test as a whole; (2) item level, which considers how well the empirical evidence matches the intention for each item (or item type); and (3) person level, which considers how well the empirical evidence matches the intention for the persons taking the test (in this case the students). The validity matrix has a separate row for each of these three, and if there were more interesting features in the situation, such as, say, raters, they would have a separate row, too. Thus, each of these is examined, in turn.

Test Level

Considering the FLP project, and also considering the fourth row of the evidence matrix, clearly the Wright map embodies crucial information. As noted above, Figure 4.3 shows one view of the Wright map resulting from the analyses carried out by the FLP project. Focus first on column 2 and row 4 of the evidence matrix—this is, in many cases, the most crucial sort of evidence that the constructs have successfully measured. This particular map also displays how the Wright map can be used formatively in the test-development process: it shows on the right-hand side the locations of the thresholds for 8 of the 12 items in the test (the 4 that are not displayed were all similar to the results for the "Equivalent Functions" item). Note that in Figure 4.3 only three levels have been denoted by different levels of shading—this is because the results showed that the locations of the thresholds within each of the EA and R levels (recall that the meanings of these abbreviations are given in Figure 4.2), the MS and US levels, and the PS and PA levels, were not distinct. Hence Wilmot and her colleagues (2011) concluded that, although the six levels could be distinguished educationally, these distinctions were not being confirmed empirically at the six-level grain-size. However, a clearer pattern results when considering just the three-level grain-size, as indicated in Figure 4.3. As an example of interpretation of these bands, we can see that, as mentioned above, most students are falling within the middle two levels (US and MS).

In addition, one can consider the overall dispersion of the item thresholds (i.e., row 4 and column 4 of the evidence matrix)[8]—beyond the considerations in the paragraph above, Figure 4.3 shows these to be well-distributed across the three (final) levels of the FLP construct. In contrast, this distribution can sometimes show sparseness of items with respect to certain levels, which would almost always be a consideration for augmenting the item set (or perhaps the set of item categories, depending on the context).

The last cell in this row refers to two issues. The first is evidence of item fit to the measurement model—a very important basis for using any of the empirical results. In this case, the fit results were not reported in the Wilmot et al. (2011) journal article, but they were reported in the dissertation that was the source of the journal paper (Wilmot, 2008, pp. 108–109). They were deleted from the Wilmot et al. (2011) paper under pressure from the journal editors, who did not see that this information was worth the page space it occupied![9] A table in that document (specifically, Table 9 in Wilmot (2008)) shows that all 12 items were found to fit the statistical model at levels considered usually acceptable. The second issue is the issue of the reliability of the Functions variable—specifically the reliability of the person estimates. In this case, the separation reliability for Form E of the test was found to be .70 (Wilmot et al., 2008, p. 273), and as this form includes 6 of the 12 open-ended items, the Spearman-Brown formula indicates that there would be a reliability of approximately .82 for the full set of 12. Of course, this would need to be considered from the point of feasibility—to gain this information, it would presumably require twice as much time, and this may not be readily available.

Item Level

Considering the FLP project, and also considering the fifth row of the evidence matrix, the fourth cell repeats the examination in the previous row (i.e., the test level), but does so for *each item*. Looking again at Figure 4.3, one can see that, although the boundaries of the bands for most of the items have successfully separated the three levels in a consistent way, this is not true for three of the items. Specifically, for the Hexagon Pattern Task, the R and EA thresholds are lower (easier) than they should be, and for the Gender Gap and Staircase Toothpick Tasks, the thresholds are generally high (harder) than they should be. Thus, a decision is required, either (1) to revise these items or generate new ones that do indeed conform to the banding as developed by the project or (2) to tolerate these discrepancies under the assumption that the weight of the item evidence supports the banding, and interpretation of student locations will not suffer unduly.

A further interesting result is referred to in the fifth cell in this row (i.e., Internal Structure: Item Level)—for each item, the examination of the mean locations of the students who were at each level of the construct map. In this case, the results for a sample of three items are shown in Table 4.1. The expected pattern is that, for each successive construct map level, the mean locations should increase along the construct, and this is confirmed for these three items. Were this not the case, then the categories that were not conforming could be useful as formative information in revising the item. In results such as these, an issue that can arise is "When is a difference large enough?" Guidance here can be derived from considering the standard errors for the students at these locations, moderated by the number of cases, as is usual for a mean.

Person Level

Considering the FLP project, and also considering the sixth row of the evidence matrix, we can see in Figure 4.3 that the distribution of the students along the variable is approximately unimodal, and that the upper tail is shorter than the lower one. In this case, this was the first time that data about this specific variable had been collected, so there were no strong expectations about what the overall pattern of the student

Table 4.1 Average Student Locations Across Levels of the Functions Variable

Item	Levels on the Functions Construct Map						
Name	OT/DK	PA	PS	US	MS	R	EA
100-Years Crude Oil	−1.31	−.46	.02	.61	.86	1.22	1.48
Hexagon Pattern	−.63	−.47	−.02	.16	.68	.88	1.23
Staircase Toothpick	.08	.50	.61	1.17	1.37	N/A	N/A

distribution might be, so no conclusions should be drawn here. But, in other situations, there may be strong expectations, such as there being a relatively uniform distribution, or that one tail is heavier than the other, etc., and this should be examined.

Considering the third cell in this row, referring to the match of the student distribution to the item distribution, one can see that quite a few the item thresholds are well above the highest estimates for the students. Again, as there was no strong expectation about this relationship in the FLP project, this is not a matter that requires a solution. However, a general design issue is involved here—as the most accurate estimation will most often occur when items and persons are located at about the same location. Thus, with the pattern observed here, we would see that the estimation of the locations of higher-difficulty items thresholds is less accurate (i.e., higher standard errors) than would be the case if there were more students at the higher levels. This might lead one to doubt some of the results for these higher thresholds, and consequently, one might question whether the boundary between the two upper bands was appropriately located. This would likely lead to a wish to gather more data on this test, especially data that came from a group of higher-ability students (e.g., some first-year college students).

The final cell in this row refers to the examination of person fit. The FLP project did not examine person fit results on from the FLP test (or, at least, did not report them). This is a somewhat surprising oversight, especially as the project devoted a great deal of effort to the gathering and analysis of individual student data in terms of think-alouds, which reflect a considerable interest in the individual person level of outcomes. An examination of person fit is important in two senses:

1. At the overall level, a finding of extensive person misfit (even in the presence of well-fitting items) is an indicator that the students who are responding to the items are not behaving in a way that is consistent with the probabilistic assumptions of the measurement model, and this deserves close attention.
2. At the individual level, it may be important to identify students for whom the test items do not seem to be operating as they usually do— such students deserve to have more data gathered on them to see if they test has not performed as well as it should.

Other Levels

Although not included in the version of the design validity evidence matrix shown in Figure 4.4, other subparts of the internal structure strand may be present in particular circumstances. One that is typical in situations where there are constructed responses is the effects that differing raters might have on the results. In fact, this is the case for the FLP tasks, and hence evidence about this was also considered. The rater can be included as an additional facet of the measurement model, and hence an additional row of the matrix could be added to help coordinate validity evidence. However, as there were just two raters in this case, this level of sophistication was not deemed necessary, and rater consistency was examined by the traditional method of correlating the raw scores of the two raters across the different items, for a sample of the students. This was reported in Wilmot et al. (2011, p. 273), where lower correlations were found for two of the items. Clearly, internal structure evidence can provide multiple ways and levels to investigate how the empirical evidence supports (or does not support) the structures and patterns that are inherent in the construct map and the items design.

Evidence Based on Relations to Other Variables

This type of evidence depends on having prior knowledge about the construct and its relationship to an "external variable." This external variable usually comes in two types: (1) there is a theory (that is believed to be correct) that says the construct should have a certain relationship to the external variable and (2) the external variable constitutes another (valid) way to measure the construct. Typical examples of these second type of external variables are: (1) clinical judgments, records, and self-reports (for psychological and health variables); (2) scores on other tests and teacher ratings and grades (for educational achievement tests); and (3) supervisor ratings and performance indicators (for business-related measures). Another source of external variables are treatment studies—where the measurer has good evidence that a treatment does indeed affect the construct, then the contrast between a treatment and a control group can be used as an external variable. Note that the relationship predicted from theory may be positive, negative, or null—that is, it is equally as important that the instrument be supported by evidence that it is measuring what it should measure (convergent evidence, which may be positive or negative depending on the way the variables are scored), as it is that it be supported by evidence that it is not measuring what it should not (divergent evidence, which would be indicated by a null relationship).

A common indicator of this relationship is the Pearson correlation coefficient between the respondents' estimated locations and the external variable. When the external variable is a dichotomy (i.e., it divides the sample into two groups), the relationship can be examined by considering the difference between the means for the two groups, and can be tested using a t-test. Both of these methods should also be employed alongside a t-test of statistical significance of a null relationship (a = .05 is the usual scientific standard). In addition to calculating this quantitative indicator, is also useful to plot the relationship, and the size of the correlation coefficient can be interpreted as an effect size. This relationship can also be considered at the item level rather than the test level (by using item scores). This gives a finer-grain examination of such evidence, though it will be likely to be more varied than the test-level results, and hence less interpretable.

The FLP project was mainly focused on test development, and the evidence gathered did not focus on its relationship with other variables. Absent an equivalent alternative test, to gather evidence in this case one might have looked to gathering teacher judgments on students' abilities on this topic.

Evidence related to other variables can help provide an "outside view" of the functioning of the instrument (as opposed to the "inside view" provided by internal; structure evidence). This can be formative when the predictions turn out to be wrong, and confirmatory when they turn out to be correct.

Evidence Based on the Consequences of Using an Instrument

Regardless of all the other evidence gathered in the above categories, if a general category of use of a particular instrument is found to have negative consequences, then that is an overriding consideration that should be taken into account when deciding whether or not to use an instrument. Note that to determine whether it is indeed the particular instrument that was causing the problem, one would have to use an alternative instrument in the same place and get different results. If it was the case, for instance, that any instrument so-used would have similar negative consequences, then the problem lies with the use of the construct, not the instrument per se. Putting that comment aside, this category of validity evidence should be seen as the real-world "complement" of all the rest. Each validity argument composed of the various subarguments from the categories above can only be a partial investigation of the very large (probably infinite) number of potential threats to validity. And this final category gives a (self-)critic the right to "20/20 hindsight" in deciding that good intentions and thorough methodology may sometimes be insufficient. See the discussion on this topic in Linn (1997), Mehrens (1997), Popham (1997), and Shepard (1997).

Because there are always important consequences of using the instruments that measurers develop (otherwise they would not have developed them), it is important to maintain a constant monitoring of good usage and positive outcomes. One might say that this is a constant in any area where products are designed, but it is particularly so in this area of measurement. Consider an analogy—when an engineer builds a bridge, and the bridge falls down, everyone can tell that there has been a failure. But in measuring students' achievement, it can be very hard for anyone to tell that "the bridge has fallen down." Often the instruments are the only basis for decisions, and may be the only data that are regularly collected, so finding evidence of problems can be quite difficult. Often they are the basis of professional standing, so that people's reputations are in danger if the instruments are criticized (this comment applies both to people who are judged using the instruments, and those who use them profession-ally). Thus, an essential component of sound measurement is a vigilant attention to consequences.

There are a couple of caveats to the position outlined in the para-graph above. First, consider the situation where a user goes against the recommendations of the instrument designers. This is a case of *abuse* of an instrument, and hence this is a problem associated with the (ab) users, not necessarily the instrument itself. Of course, if the designers do not provide sufficient information on what they consider to be appropri-ate usage, then they have indeed promoted poor consequences.[10] Second, positive consequences of particular types of usage should not be general-ized to other types of usage (say, where a test works out well as a high school grade assessment, that does not mean it will be justified to use it as a college-readiness measure): Equally, evidence of negative consequences for particular types of usage should not be generalized to other types (say, where a test is found to be poor at diagnosis, that does not mean it should not be considered for another usage, such as summative assessment).

Considering the FLP project, and also considering the eighth row of the evidence matrix, the investigators in the FLP project devised an inter-esting and innovative evaluation of (certain) potential outcomes of the use of their test associated with the last cell in the row—they designed a method of reporting student outcomes at a teacher level, and then tried this out on some teachers. An example of the report is shown in Fig-ure 4.5 (Wilmot et al., 2011, Appendix D). This report shows (on the left-hand side) the distribution of student estimates for the students in a teacher's four classes, and (on the right-hand side) the means for the students from each of those classes. The results of the study are quite complex, reflecting both teacher understandings and the context of the data situation, so it is not possible to report a single outcome. However, in summary Wilmot et al. reported that the reports "proved useful as a formative measure to help them check student progress, view learning as a trajectory, plan curriculum across years and make instructional and curricular decisions at all levels of the system" (2011, p. 281).

Consequences are, in some sense, the whole point of making an instru-ment, so paying attention here is crucial. Unfortunately, developers often disappear once the instrument is released, leaving no one with the impor-tant role of checking-up.

Fairness

Instruments are typically used across a wide spectrum of respondents— one important requirement of fairness is that, across important subgroups,

Estimated Average Proficiency in Math Courses
(Lower/Upper range of 95% Confidence Interval)

Scaled Score	Distribution of Students (n = 47)	Grade Confidence Interval	STUDENTS ARE ACTIVELY LEARNING....
3 –			The SIGNIFICANCE of the connections and how to APPLY them or GENERALIZE.
–			
–			
–			
2 –			
–			
–	X		
–	XX		
1 –			How to MAKE CONNECTIONS across more than one representation.
–	XXX		
–	XXXXX	8th grade (–.34, .26)	
–	XXX		
0 –	XXXXXXXXX		
–			
–	– – – – – – – X	7th grade (–1.06, –.46)	
–	XXXXXXXXX	9th grade (–1.32, –.03)	
–1 –	X		The PREREQUISITE SKILLS to make connections across representations of functions.
–	XXXXXXX		
–	X	6th grade (–.98, –.60)	
–	X		
–2 –	XX		
–			
–	X		
–			
–3 –	X		

Each X represents 1 student.
Dashed line (–) represents average student proficiency.

Figure 4.5 A teacher report for the Functions variable.

items function in a similar way for respondents who are at the same location—that is, they should exhibit no evidence of *differential item functioning* (DIF).[11] Here one is focusing on whether the items in the instrument behave in a reasonably similar way across different subgroups within the sample. Typically these subgroups are gender, ethnic, or socio-economic groups, although other groupings may be relevant under particular circumstances. The technique can be applied to other subgroups, such as those identified in investigative studies (e.g., respondents who use different cognitive strategies), or could even be composed of respondents with different sets of scores on the test—while these may be interesting for a variety of reasons, they are not specifically the focus of this section (AERA et al., 2014).

First, a bit of jargon. If the responses to an item have different frequencies for different subgroups, then that is evidence of *differential impact* of the item on those subgroups. Although such results may well be of interest for other reasons, they are not generally the focus of DIF studies. Instead, DIF studies focus on whether respondents at the same locations give similar responses across the different subgroups. Thus, in the FLP example, if two subgroups gave different rates of correct response to a particular item, say that students from one school gave different responses from those in another school, then that would constitute differential impact on the results of the measurement, and could be an interesting result in itself. But the issue of DIF would not necessarily be raised by such a result—it could be that the students in the first school had had a different type of instruction than the students in the second school. To make this distinction between differential impact and DIF, the measurer needs to control the comparison for the different levels of FLP variable between the two groups. Or, in other words, the measurer needs to compare the item responses for students at the same location in the two groups. Thus, in the example, one would want to know if respondents from the two groups *at the same location on the variable*, were giving similar responses or not.

Moving down to the final row of the evidence matrix, there are several different techniques available for investigating DIF, among them techniques based on linear and logistic regression, and techniques based on log-linear models (see Holland & Wainer, 1993, for an overview). One technique is based on the Rasch measurement model. In summary, calibrate the FLP variable along with an interaction between the student ability and the group indicator for the two subgroups, and then examine the resulting group interaction parameter estimates for important and statistically significant effects. Such differences will indicate some form of DIF (Wilson, 2004). Using a technique equivalent to this, Wilmot (2008) reports that, among the four items that had sufficient data to allow such a comparison, two were in the "negligible" category, and the other two were in the "medium" category. According to standard interpretation (Wilson, 2004), these results are not considered grounds for concern.

Fairness has aspects that occur in all the other aspects described above. We have focused here on the item level (using DIF), but actually, the possibilities range much more widely than that, from differences in internal structure between different groups, to issues about wording and tone in writing items.

Summary and Conclusion

The quest for validity for a test's outcomes and interpretations is a never-ending task—of course, there is no truly "valid" test. In fact, it has become a commonplace to say that, instead, there can only be validity for particular usages of a test. In this chapter, we have sketched out a middle ground to this, where evidence is gathered relating to the theory and materials developed to establish a case for broad confidence in a range of possible usages. Of course, the usage in a particular context would then have to be investigated for the "local" validity of that usage, say, along the lines described by Kane (2006)—there are many ways that one could design a case for the validity of a particular usage.

What has been put forward in this chapter is that the validity argument must be tethered to the theories and materials that were used to design the test, and that when those theories and materials are richer then the argument can be stronger. Kane's distinction between the interpretive argument and the validity argument has been used as a perspective on this: The BEAR Assessment System (BAS) was used as a framework to establish the interpretive argument, and this gave a logic from which to gather validity evidence that spans the six strands of validity evidence from the *Standards* (AERA et al., 2014). The validity matrix organizes the way that the strands of the validity argument can be based on the different forms of evidence that can be collected as the measurer passes through the four building blocks. As noted above, it is not always that case that journal editors are aware of the importance of some aspects of this validity evidence, but hopefully, a chapter such as this can be used to help educate such editors (and reviewers). The framework of the BAS plus the *Standards* provides researchers with a way to develop and report on the validity evidence and ensuing argument.

Of course, there are other such framings, such as "evidence centered design" (ECD; Mislevy, 1996), and indeed, there are similarities between this and BAS. Following a design approach such as the one described in this chapter, or ECD, will allow the measurer to create a strong and comprehensive plan to gather validity evidence.

The account given above is a sketch of the many possible tactics one could deploy in using the BAS to develop a validity argument, no more than can fit into a single chapter. It might be seen that the tactics reported above are limited to simple unidimensional testing contexts, but, indeed, this would be a misunderstanding based on the choice of example—there is no restriction to single-variable contexts, as a single test may be used to assess multiple constructs, and the types of validity evidence described above could be gathered for each construct, as well as validity evidence related to the ensemble of constructs.

If a mathematics educator or mathematics education researcher wanted to use this new framework, then the best way to do so would be to (1) design and develop an instrument using the BAS, and (2) design and carry out a validity evidence–gathering exercise for that instrument. One could start with a preexisting instrument, but this will often prove weak in the very first building block, the construct map, or its equivalent, which makes the investigation of the rest not so interesting. Thus, to avoid this frustration, we advise following the BAS approach to development (Wilson, 2004), and the framework described here for evidence gathering.

Acknowledgments

The original data collection in the FLP study described in this chapter was funded through the Berkeley Futures Project (BFP), University of California. We would like to thank the editors and two blind reviewers for their helpful and insightful comments. Any errors or omissions remain the responsibility of the authors

Notes

1. Adapted from a Keynote Speech presented at the Validity Evidence for Measurement in Mathematics Education (V-M^{2}Ed) Conference, San Antonio, TX, April 2–3, 2017.
2. Note that this is somewhat unusual, as most learning progressions are conceived as being more complex See Wilson (2009) for a discussion about how the simple example shown here can be expanded to multiple dimensions, and branching, while still maintaining the essential nature of the concepts discussed here.
3. The analyses for these Wright maps were performed using the *ConQuest* software program (Adams, Wu, & Wilson, 2018), which implements an EM algorithm for estimation of multidimensional Rasch-type models.
4. Counting the column headings as the first row.
5. Counting the row headings as the first column.
6. Note that in some contexts only one of the third and fourth cells might be applicable, depending on whether the context is focused on items, or categories within levels, or both.
7. Note that these do not have to take place in an actual lab, but can be conducted anywhere that is quiet and not distracting for the student.
8. Note that only the fourth row is relevant here, not the second, as the model is based on item thresholds only, not item difficulty parameters.
9. This unfortunate event illustrates the importance of connecting the validity evidence collected to the validity argument advanced. Having all of the

right evidence on hand doesn't help much if we can't show our readers how it connects and why it's important (Personal communication, chapter reviewer).

10. As a caveat to that caveat, if a vendor sells an instrument that they know is being abused, then that does not relieve them of their obligations to monitor consequences and take action to correct negative ones (Personal communication, chapter reviewer).
11. Where there is a sound alternative test, one can also examine DTF—differential test functioning—but this is rarely available, so we will not discuss it here.

References

Adams, R. J., Wilson, M., & Wang, W. (1997). The multidimensional random coefficients multinomial logit model. *Applied Psychological Measurement*, 21(1), 1–23.

Adams, R. J., Wilson, M., & Wu, M. (1997). Multilevel item response models: An approach to errors in variables regression. *Journal of Educational and Behavioral Statistics*, 22(1), 47–76.

Adams, R. J., Wu, M., & Wilson, M. (2018). ACER ConQuest. In W. van der Linden (Ed.), *Handbook of item response theory, volume three: Statistical tools, and applications*. New York, NY: Chapman & Hall.

American Educational Research Association, American Psychological Association, & National Council on Measurement in Education (AERA, APA, & NCME). (1985). *Standards for educational and psychological testing*. Washington, DC: Authors.

American Educational Research Association, American Psychological Association, & National Council on Measurement in Education. (AERA, APA, & NCME). (2014). *Standards for educational and psychological testing*. Washington, DC: Authors.

Biggs, J. B. (1999). *Teaching for quality learning at university*. Buckingham, UK: SRHE & Open University Press.

Biggs, J. B., & Collis, K. F. (1982). *Evaluating the quality of learning: The SOLO taxonomy*. New York: Academic Press.

California State University Academic Senate. (1997). *Statement of competencies in mathematics expected of entering college students*. Long Beach, CA: Author.

Common Core State Standards Initiative. (2010). *Common core standards for mathematics*. Washington, DC: National Governors Association Center for Best Practices and Council of Chief State School Officers.

Holland, P. W., & Wainer, H. (1993). *Differential item functioning*. Hillsdale, NJ: Lawrence Erlbaum.

Kane, M. (2006). Validation. In R. Brennan (Ed.), *Educational measurement* (4th ed., pp. 17–64). Westport, CT: American Council on Education and Praeger.

Leinhardt, G., Zaslavsky, O., & Stein, M. K. (1990). Functions, graphs, and graphing: Tasks, learning, and teaching. *Review of Educational Research*, 60, 1–64.

Linn, R. L. (1997). Evaluating the validity of assessments: The consequences of use. *Educational Measurement: Issues and Practice*, 16(2), 5–8.

Mehrens, W. A. (1997). The consequences of consequential validity. *Educational Measurement: Issues and Practice*, 16(2), 5–8.

Mislevy, R. J. (1996). Test theory reconceived. *Journal of Educational Measurement, 33*, 379–416.

Moschkovich, J., Schoenfeld, A., & Arcavi, A. (1993). Aspects of understanding: On multiple perspectives and representations of linear relations and connections among them. In T. Romberg, T. Kennedy, & E. Fennema (Eds.), *Integrating research on the graphical representations of functions* (pp. 69–100). Hillsdale, NJ: Lawrence Erlbaum.

National Council of Teachers of Mathematics. (2000). *Principles and standards for school mathematics*. Reston, VA: Author.

National Research Council (NRC). (2001). *Knowing what students know: The science and design of educational assessment*. Committee on the Foundations of Assessment. In J. Pellegrino, N. Chudowsky, & R. Glaser (Eds.), *Division on behavioral and social sciences and education*. Washington, DC: National Academy Press.

Nisbett, R., & Wilson, T. (1977). Telling more that we can know: Verbal reports on mental processes. *Psychological Review, 84*, 231–259.

Piaget, J., Grize, J.-B., Szeminska, A., & Bang, V. (1977). *Epistemology and psychology of functions*. Dordrecht: D. Reidel Publishing.

Popham, W. J. (1997). Consequential validity: Right concern-wrong concept. *Educational Measurement: Issues and Practice, 16*(2), 9–13.

Rasch, G. (1960/1980). *Probabilistic models for some intelligence and attainment tests*. Chicago: University of Chicago Press (original work published 1960).

Shepard, L. A. (1997). The centrality of test use and consequences for test validity. *Educational Measurement: Issues and Practice, 16*(2), 5–8.

Wilmot, D. B. (2008). *Assessing progress toward college readiness with psychometric and cognitive models of student learning in mathematics* (Doctoral dissertation). Berkeley: University of California.

Wilmot, D. B., Schoenfeld, A., Wilson, M., Champney, D., & Zahner, W. (2011). Validating a learning progression in mathematical functions for college readiness. *Mathematical Thinking and Learning, 13*(4), 259–291.

Wilson, M. (2004). *Constructing measures: An item response modeling approach*. Mahwah, NJ: Erlbaum.

Wilson, M. (2009). Measuring progressions: Assessment structures underlying a learning progression. *Journal for Research in Science Teaching, 46*(6), 716–730.

Appendix

1. For the following geometric pattern, there is a chain of regular hexagons (meaning all 6 sides are equal):

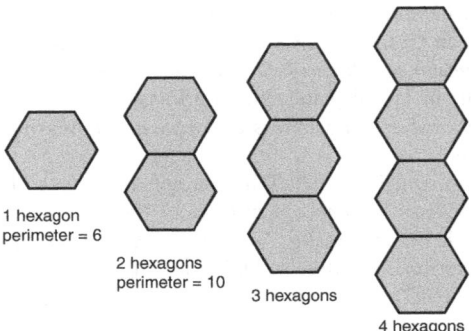

1a.) Complete the table showing the number of hexagons in a chain and the perimeter (number of outside edges).

NUMBER OF HEXAGONS	PERIMETER
1	6
2	10
3	
4	
5	

1b.) Describe the process for determining the perimeter for 100 hexagons, without knowing the perimeter for 99 hexagons.

_____

_____

_____

_____

_____

_____

_____

_____

1c.) Write a formula to describe the perimeter for any number of hexagons in the chain (it does not need to be simplified).

_____

_____

Figure 4.A An example item from the FLP test (the Hexagon Pattern Task).

1d.) Explain why you think your formula in (1c) is correct.

_____

_____

1e.) Suppose instead of regular hexagons, you form a chain with regular
n-gons (where *n* is any number of sides). Describe the process for
determining the perimeter for any number of regular *n*-gons in a chain.

_____

_____

_____

_____

_____

1f.) Write a formula to describe the perimeter for any number of regular
n-gons in a chain (it does not need to be simplified).

_____

_____

1g.) Explain why you think your formula in (1f) is correct.

_____

_____

5 Validity Arguments for Instruments That Measure Mathematics Teaching Practices

Comparing the M-Scan and IPL-M

Temple A. Walkowiak, Elizabeth L. Adams, and Robert Q. Berry

Given the increasing desire to use multiple measures to document what happens inside classrooms, instruments that measure teaching practice are becoming more widely available (Cohen & Goldhaber, 2016). The uses of these instruments (e.g., coaching, evaluation, research) directly inform policy decisions and research implications (Jacobs & Spangler, 2017). Thus, the information generated with these instruments needs to be valid, reliable, and appropriate for the intended purpose (Kane, 2006; O'Leary, Hattie, & Griffin, 2017). Developers have a critical responsibility to evaluate an instrument's strengths and weaknesses in relation to the intended purpose. However, challenges exist when examining validity evidence for instruments that measure teaching practice. Measurement experts have written about collecting validity evidence for traditional assessments (e.g., achievement tests) for decades. As a result, widely accepted frameworks exist for validation (American Educational Research Association, American Psychological Association, & National Council on Measurement in Education [AERA, APA, & NCME], 2014; Kane, 2013). On the other hand, collecting validity evidence for instruments that measure teaching practice is relatively new and less straightforward (Bell et al., 2012). These instruments can be informed by multiple perspectives including those of the teacher, students, parents, and external evaluators. For example, teaching practice might be measured with an observation, portfolio, student or parent survey, student test scores, or teacher instructional logs. The types of validity for each of these instruments may differ in some cases substantially. In this chapter, we compare the collection of validity evidence for two instruments that measure teaching practice, an observational protocol and an instructional log.

Validity is the most fundamental consideration in developing and evaluating measures because it examines the degree to which the data generated are supported by their interpretations and uses (AERA et al., 2014). According to Messick (1995), validity is a multifaceted, but unified concept focused on evidence to evaluate the intended inferences about a specific construct, defined as teaching practice in this chapter. As examples, validity evidence examines: (1) the actual language or content of the instrument, (2) the thought or cognitive processes in which participants engage as they complete it, and (3) the theoretical and empirical relationships amongst the items. More specifically, the *Standards for Educational and Psychological Testing* state the need to develop a coherent validity argument with an emphasis on five sources of evidence (AERA et al., 2014):

- Content: The alignment between an instrument's content and the construct it is intended to measure;
- Response process: The fit between the construct and the cognitive processes used by those assigning scores;
- Internal structure: The degree to which the relationships amongst items and components of the instrument conform to the construct;
- Relations to other variables: Analyses of the relationships between generated data and variables external to the instrument; and
- Consequences: The intended uses of the generated data are justified and beneficial for users.

The *Standards* express that the validity of the score interpretations depends on producing and integrating these types of evidence. Several evaluations of instruments that measure teaching practice have organized their validity argument around these types of evidence (e.g., Adams et al., 2017; Camburn & Barnes, 2004; Kurz, Elliott, Kettler, & Yel, 2014; Walkowiak, Berry, Meyer, Rimm-Kaufman, & Ottmar, 2014).

Kane (2006) presented a framework that complements the *Standards* and focuses on evaluating the validity of the *score interpretations*, referred to as the interpretation-use argument (IUA). Kane (2013) states that the IUA "specifies the claims (the inferences and assumptions inherent in the proposed interpretations and uses of test scores) that are to be evaluated in the validation effort" (p. 451). Kane (2013) outlines four common inferences to be evaluated within an IUA: scoring, generalization, extrapolation, and score use or implication. Both the *Standards* and Kane (2013) also emphasize the importance of score reliability, which ensures that scores are minimally influenced by extraneous sources other than the target construct. As noted in the *Standards*, "the level of reliability/ precision in scores has implications for validity" (p. 34).

Our Purpose

We have two main purposes for this chapter. Our first purpose is to describe each of Kane's inferences in the context of *measuring mathematics teaching practices* with two different instruments, the Mathematics

Scan (M-Scan; Berry et al., 2013; Walkowiak et al., 2014) and the Instructional Practices Log in Mathematics (IPL-M; Walkowiak & Lee, 2013; Walkowiak, Adams, Porter, Lee, & McEachin, 2018). Specifically, we present each inference and its accompanying set of assumptions for each instrument. A second purpose of this chapter is to *pair Kane's argument-based approach with the validity evidence types* presented in the *Standards* in our description of how we evaluated or would evaluate the validity of the inferences and assumptions. By overlaying these two frameworks, we provide a model for organizing validity arguments in a comprehensive, yet systematic way. To our knowledge, the overlaying of these frameworks is a recent phenomenon. Furthermore, our work focuses on the measurement of teaching practices, which is often considered complex in terms of examining validity (Bell et al., 2012).

Measuring Mathematics Teaching Practice

Teaching practice extends beyond lesson implementation to teachers' planning, interactions with curricula, and analyses of student assessment data. However, we focus on measuring the teaching practices utilized during lesson implementation. By teaching practice, we mean the instructional choices and moves that a teacher makes. Students' opportunities to learn are a direct result of a teacher's instructional practices (Elliott, 2015). As such, we also examine student behaviors as fitting under the umbrella of teaching practice. Here, by "student behaviors," we mean the processes or practices that students utilize as it relates to the mathematical content (e.g., math talk, representations). Therefore, our goal is to measure teachers' instructional practices and concurrent student behaviors that occur during a mathematics lesson. This type of measurement has implications for several types of users including:

- Researchers who are interested in understanding the *relationships* between teaching practice and other constructs such as teacher content knowledge or student characteristics (e.g., achievement, demographics),
- Teacher educators and program evaluators who need information that will inform *program improvements*,
- School leaders and policymakers who need information that will inform *school improvement*, and
- Teachers who are interested in identifying their own *strengths and opportunities for growth*.

Given the critical need to measure teaching practices (Jacobs & Spangler, 2017), several instruments have emerged, but we focus our discussion here on observational protocols and instructional logs. First, observational measures have been considered the "gold standard" in measuring teaching practice for over a decade (Borko, Stecher, Alonzo,

Moncure, & McClam, 2005). Observations provide information about teaching practice that is directly observed by a rater. However, the validity of observation scores depends on rater reliability (Meyer, Cash, & Mashburn, 2011), and conducting observations is resource and time intensive (Praetorius, Pauli, Reusser, Rakoczy, & Klieme, 2014). Despite these challenges, observational protocols have been utilized to measure teaching practices broadly (Pianta, La Paro, & Hamre, 2008; Danielson, 2013) and specific to mathematics (Hill et al., 2008; Walkington et al., 2012, 2014).

A second instrument is daily teacher instructional logs, which have shown promise for measuring teaching practice (Kurz et al., 2014; Rowan, Harrison, & Hayes, 2004). Instructional logs are usually formatted as a checklist or Likert-type items asking teachers the extent to which they or their students engaged in certain behaviors during that day's lesson. Instructional logs are attractive because they can be completed by teachers across several instructional days at several points of time across the school year. Thus, logs lend themselves to multiple observations, which may contribute to increased stability in scores when compared to observations. Because they can be completed electronically, fewer resources are necessary to collect instructional log data compared to observational data. However, instructional logs have limitations for those interested in accountability decisions such as school leaders and policymakers. Because teachers report their own practices, responses are more likely to include bias if teachers know the information will be used to evaluate them or their schools. Despite these limitations, evidence suggests that daily instructional logs are a viable tool for measuring teaching practice in mathematics and language arts (Rowan & Correnti, 2009; Rowan et al., 2004) and science (Adams et al., 2017; Hayes, Lee, DiStefano, O'Connor, & Seitz, 2016) to inform research and program evaluation.

It is important to note a critical difference in the aspects of teaching practice that an observational protocol and an instructional log measure. Observational protocols provide information about observed teacher and student behaviors, which leads to inferences about the *quality* of students' opportunities to learn. Instructional logs provide information about teacher-reported teacher and student behaviors, which lends to inferences about the *frequency and duration* of students' opportunities to learn, not the quality of those opportunities. This is a critical difference in the inferences that can be made based on the data collected with these two instruments.

The Two Focal Measures: M-Scan and IPL-M

When a researcher is deciding on whether to utilize the M-Scan, the IPL-M, or both, the decision depends upon the researchers' goals and how they have operationalized students' "opportunities to learn," as described in the previous paragraph. Table 5.1 presents a comparison of the two focal instruments. Broadly, the M-Scan measures the extent to which instruction includes standards-based mathematics teaching practices. These teaching practices are characterized by students' opportunities to engage in mathematical behaviors outlined by standards documents (National Council of Teachers of Mathematics [NCTM], 2000; National Governors Association Center for Best Practices & Council of Chief State School Officers [NGA CBP & CCSSO], 2010). Specifically, the M-Scan was designed to measure nine dimensions of mathematics instruction (presented in Table 5.1), most of which were previously identified by Borko and colleagues (2005). The M-Scan was initially designed to examine elementary mathematics instruction, but we hypothesize that the rubrics would also produce valid data for secondary instruction, a hypothesis that needs to be tested. During the development of the M-Scan, researchers developed coding rubrics for each dimension and used existing theory to map the dimensions onto four domains: tasks, discourse, representations, and coherence. Using coding rubrics, the M-Scan's dimensions are coded on a scale of 1 to 7, modeled structurally after the scoring protocol for the Classroom Assessment Scoring System (CLASS; Pianta et al., 2008) with descriptors for low (1 and 2), medium (3, 4, and 5), and high levels (6 and 7). Validity evidence has been collected for dimension-level scores, which will be presented in this chapter; for a complete and thorough description of evidence of validity and score reliability, see Walkowiak et al. (2014).

The IPL-M, a teacher-reported log of instructional practices, measures the frequency and duration (as a proportion of time) of teachers' instructional practices and signals the types of learning opportunities provided to students during mathematics instruction. Specifically, the items are grouped into five scales that measure students' opportunities to engage in *Problem Solving*, make *Connections*, engage in *Math Talk*, use *Multiple Representations*, and utilize *Mathematical Procedures*. The IPL-M was initially designed to measure instruction in elementary classrooms. After participating in a training on how to interpret the log items, teachers complete the daily log for multiple lessons across the school year (in our large-scale implementation of the log, teachers logged for three 15-day time points). Focusing on one specific class of students, a teacher documents the behaviors in which at least half of the students engaged during the mathematics lesson by responding to a prompt stating, "During today's lesson, how much time did students in the target class _____?" The blank is filled in with 30 different items such as "demonstrate different ways to solve a problem" or "discuss ideas, problems, solutions, or methods in large group." The teacher responds on a 4-point scale to indicate how much time the students in the target class engaged in the given behavior: not today (not done

Table 5.1 A Comparison of the M-Scan and the IPL-M

	M-Scan	IPL-M
What is it?	Observational measure	Daily teacher instructional log
What does it measure?	Standards-based mathematics teaching practices	Frequency and duration of students' learning opportunities during mathematics lessons
Which instructional dimensions are included? *For M-Scan only, specifies the theoretical domain to which the dimension maps	*Dimensions (Domain*):* Cognitive Demand (Tasks) Problem Solving (Tasks) Connections and Applications (Tasks) Student Use of Representations (Representations) Teacher Use of Representations (Representations) Use of Mathematical Tools (Representations) Math Discourse Community (Discourse) Explanation and Justification (Discourse) Structure of the Lesson (Coherence) Mathematical Accuracy (Coherence)	*Dimensions:* Problem Solving Connections Math Talk Multiple Representations Mathematical Procedures
How many indicators or items per dimension?	*Indicators per Dimension:* 2–3	*Items per Dimension:* 5–9
How is it completed?	Coders use a scoring rubric with indicators to assess each dimension based on the observed, video-recorded lesson	Teachers respond to the following prompt: "During today's math lesson, how much time did students in the target class _____?"
When is it completed?	Coders complete the M-Scan immediately after watching a video-recorded lesson	Teachers complete the log electronically on the same day the lesson was taught
What is the numerical scale?	Scale of 1–7: 1–2 → Low 3–5 → Medium 6–7 → High	Scale of 1–4: 1 → Not today 2→ Little 3→ Moderate 4→ Considerable
Who completes it?	Coders who participate in a 2.5-day training	Teachers who participate in a 2.5-hour training
Who might use the scores?	Researchers, professional development providers, school leaders, teachers	Researchers, professional development providers, school leaders, teachers
What is the intended use?	Research, program evaluation, teacher coaching	Research, program evaluation, teacher coaching

during today's instruction); little (made up a relatively small part of the instruction; moderate (made up a large portion but NOT the majority of instruction); and considerable (made up the majority of today's math instruction).

As previously mentioned, there are four different types of users for instruments like the M-Scan and IPL-M: researchers; teacher educators and program evaluators; school leaders and policymakers; and teachers. Both the M-Scan and IPL-M were initially developed for use by researchers to measure teaching practices, albeit in different ways, and subsequently to examine relationships between teaching practices and other constructs. Although the initial intent of both measures was for use by researchers, in this chapter we explore how the interpretation-use arguments relate to other users. It is important to note that *neither* measure was designed as a tool to evaluate an individual teacher's instruction as a part of performance evaluations completed by the principal.

IUAs for the M-Scan and IPL-M

We now present four types of inferences commonly found in IUAs (Kane, 2013) for the M-Scan and IPL-M: scoring, generalization, extrapolation, and implications. We include these four types of inferences as a comprehensive example of how one might lay out the IUA for a measure of teaching practice. Table 5.2 displays the four inferences from Kane (2013) overlaid with the evidence types from the *Standards* and highlights which evidence types support each inference for M-Scan and IPL-M. As we present the inferences and assumptions, we provide examples of data collection for each evidence type, and we highlight the differences in the nature of the validity evidence for an observational protocol versus an instructional log. For many of our examples, we have collected these

Table 5.2 Pairing Kane's (2013) Framework With Evidence Types From the *Standards* (AERA et al., 2014) for M-Scan and IPL-M

Inferences (Kane, 2013)	Measure	Evidence Types (AERA et al., 2014)					
		Test Content	Response Processes	Internal Structure	Relations to Other Variables	Reliability	Consequences
Scoring	M-Scan	X	X	X		X	
	IPL-M	X	X	X	X		
Generalization	M-Scan					X	
	IPL-M					X	
Extrapolation	M-Scan		X		X		
	IPL-M		X		X		
Implications	M-Scan		X		X		X
	IPL-M		X		X		X

evidences of validity in our own validation work (Walkowiak et al., 2018, 2014), and we describe that in this chapter. However, there are examples where we have not collected the evidence, but we include these examples for two reasons: to highlight that validation is an ongoing, iterative process (Kane, 2013) and to illustrate and inform the field about the type of validity evidence that could be collected to examine a given assumption.

The pairing of the frameworks in Table 5.2 reflects the types of evidence for the M-Scan and IPL-M. This is not to say that the types of evidence collected for each inference will be similar for other measures; in fact, it will likely differ. The purpose of Table 5.2 is to serve as one example of how types of evidence overlap with each of Kane's four inferences. Throughout the text, we refer to the developers of each measure as "we" because a subset of this chapter's authors were developers and/or involved in the validation of each measure (M-Scan: Authors 1 and 3; IPL-M: Authors 1 and 2).

The Scoring Inference

The scoring inference connects what occurred during the implementation of a mathematics lesson, either observed or reported by a teacher, with the scores given. For both M-Scan and IPL-M, the overarching issue of validity is that the scores mean as close to what was intended as possible. Specifically, we need to evaluate whether the scores are appropriate, accurate, consistent, bias-free, and aligned with the hypothesized theoretical structure. Table 5.3 presents the sets of assumptions that accompany the scoring inference; there are five assumptions for the M-Scan and four assumptions for the IPL-M. Table 5.3 also displays five types of validity evidence that evaluate these sets of assumptions; the validity evidence is organized by the evidence types presented in the *Standards* (AERA et al., 2014). The sets of assumptions for the two instruments are strikingly similar, but we highlight the differences between the nature of the validity evidence.

Assumptions MS.S.1 and IM.S.1

The first assumption for the scoring inference relates to the instrument being completed accurately, consistently, and as intended by the developers. Since the M-Scan is completed by an external coder, the coder's understanding of the rubrics and ability to apply the rubrics is critical. Similarly, for the IPL-M, the teacher's interpretation of each item must be aligned with the developers' intent. However, to support these assumptions, the types of validity evidence from the *Standards* vary for each measure: response processes, test content, and reliability for M-Scan versus response processes and relations to other variables for IPL-M (see Table 5.3).

Table 5.3 Assumptions and Validity Evidence for the Scoring Inference for M-Scan and IPL-M

Assumptions: Scoring (S)	Validity Evidence				
	Test Content	Response Processes	Internal Structure	Relations to Other Variables	Reliability
M-Scan (MS)					
MS.S.1*: Scoring with the M-Scan is applied: (a) accurately and (b) consistently. MS.S.2: Scores on the M-Scan represent the extent to which mathematics instruction is standards-based. MS.S.3: Scores on the M-Scan represent variation in teaching practices. MS.S.4: Scoring with the M-Scan is bias-free. MS.S.5: Scores on the M-Scan dimensions correlate in ways that are theoretically supported.	MS.S.1a: Examination of exact-match and within-one agreement between master codes and coders' assigned scores MS.S.2: Expert review of the dimensions and indicators on the M-Scan	MS.S.1a: Examination of coders' interpretations of the rubrics through cognitive interviews MS.S.1a: Master coder's qualitative analysis of coders' response processes AND/OR Qualitative comparison of coders' response processes to master coders' response processes MS.S.3: Examination of descriptive statistics for each dimension, expecting coders to use the full range of possible scores and expecting substantial variation (as measured by standard deviations)	MS.S.2: Confirmatory factor analysis (CFA) to empirically test the theoretical structure MS.S.3: CFA to determine if M-Scan domains explain variation between teachers MS.S.5: Bivariate correlations between M-Scan dimensions MS.S.5: CFA to examine the correlations between M-Scan domains		MS.S.1b: Examine inter-rater agreement using Cohen's Kappa coefficient MS.S.1b: Qualitative comparison of coders' response processes to one another MS.S.4: Generalizability study to determine how much score variance is explained by the coder by using coders that represent range of expertise

IPL-M (IM)	Test Content	Response Processes	Internal Structure	Relations to Other Variables	Reliability
IM.S.1**: Items on the IPL-M are completed as intended. IM.S.2: Scale scores on the IPL-M represent the frequency and duration (as a proportion of time) of students' opportunities to engage in Problem Solving, make Connections, engage in Math Talk, use Multiple Representations, and utilize Mathematical Procedures. IM.S.3: Scores on the IPL-M represent variation in teaching practices and distinguishes between teachers in terms of their instructional practices. IM.S.4: Scores on the IPL-M scales correlate in ways that are theoretically supported.	IM.S.2: Expert review to assess whether items represent the construct/scale they are intended to measure	IM.S.1: Examination of logging teachers' interpretations of items through cognitive interviews IM.S.3: Examination of item-level descriptive statistics, expecting logging teachers to use full range of scores and expecting substantial variation (as measured by standard deviations). IM.S.3: Examination of item-level intraclass correlation coefficients (ICCs) to assess if items explain between-teacher variation	IM.S.2: Two-level CFA to empirically test the factor structure IM.S.3: Two-level CFA to determine if scales explain variation between teachers IM.S.4: Two-level CFA to examine correlations between IPL-M scales	IM.S.1: Comparison of teachers' log responses to observers' log responses for same lesson (examine item-level agreement and scale-level correlations)	IM.S.2: Examination of scale-level alphas to assess scale strength

*MS.S.1 stands for M-Scan, Scoring Inference, Assumption 1; subsequent labels have same pattern.
**IM.S.1 stands for IPL-M, Scoring Inference, Assumption 1; subsequent labels have same pattern.

MS.S.1

To test that M-Scan coders apply the scoring rubrics *accurately*, we examine two types of validity evidence: test content and response processes. If the content of the scoring rubrics is sufficiently descriptive such that coders understand the differences among the various levels on the rubric, then we expect coders' assigned scores to be accurate as measured by a comparison to master codes (or codes generated by M-Scan experts). For M-Scan, coders first participate in a training on interpreting and applying the rubrics for each dimension. After training, but before coding lessons on their own, coders independently watch and code a set of 6 to 10 mathematics lessons that have been coded by master coders. The newly trained coders must obtain at least 80% within-one agreement with the master codes (e.g., if 4 is the master code, then 3 or 5 would be considered accurate). These data are examined across all codes generated for the set of lessons and across codes for an individual dimension to determine if a coder is scoring accurately before coding lessons independently. Furthermore, once a coder begins scoring a data set, there is regular monitoring to ensure drift in coding does not occur. Specifically, M-Scan coders are expected to code the same video as a master coder every two weeks at a minimum. These drift checks facilitate accuracy in coders' assigned scores.

We can also test for *accuracy* in coding by examining the response processes of coders. While we have not formally conducted cognitive interviews with M-Scan coders, these data would indicate whether coders are interpreting the rubrics as intended (Willis, 2005). After coders think aloud about how they applied the rubrics, a researcher assesses the extent to which the coders are applying the rubrics accurately or as intended by M-Scan developers. Another source of response-process evidence, in this case that we have collected, is a master coder's qualitative analysis of coders' response processes. On the M-Scan coding form, coders record comments to support the score assigned for a particular dimension. These comments were analyzed by a master coder for a set of 60 lessons to determine the extent to which the rationales were aligned with the descriptors on the coding rubrics. Overall, the analyses indicated that 87.7% of the coders' (n = 3) rationales were aligned with the coding guide descriptors, suggesting that the coders were applying the coding rubrics accurately for a vast majority of the lessons (Walkowiak et al., 2014).

The second scoring characteristic in this assumption is *consistency* across coders. For consistency, we examine two forms of evidence that we classify as reliability evidence. First, as part of our drift checks for M-Scan, teams of coders watch the same lesson (along with a master coder). We utilize these data to examine inter-rater agreement through Cohen's Kappa coefficient. The coefficient provides information to evaluate the extent to which coders who observed the same lesson agree on scores. Second, in the aforementioned set of 60 lessons where rationales for codes were assessed, the study included three different coders. We qualitatively analyzed the rationales to make sure response patterns were consistent across the 60 lessons. Examining the response patterns provided insight as to whether coders were applying the rubrics consistently (Walkowiak et al., 2014).

IM.S.1

Similar to the M-Scan, we examine evidence of response processes as evidence that the tool measures what it is intended to measure. To examine response processes of logging teachers in our study, we engaged in cognitive interviews with teachers, asking them to think aloud as they responded to the IPL-M items (Willis, 2005). We focused our analysis on how the teachers comprehended the item, retrieved the relevant information about their lesson from their memory, estimated the amount of time students engaged in the behavior defined by the item, and mapped their estimation to the four-point response scale (Tourangeau, Rips, & Rasinski, 2000). Overall, the cognitive interviews support that teachers interpret the items as intended (Walkowiak et al., 2018).

A key difference between the M-Scan and IPL-M relates to consistency in scoring. Because a teacher's perspective of her lesson as she is teaching is different than an observer's, we describe comparisons between an observer's and a teacher's log for the same lesson as evidence of relations to other variables. For a total of 28 lessons in 17 teachers' classrooms, an observer visited the classroom, observed the lesson, and then completed the log. For these pairs of logs completed by the teacher and the live observer, two analyses were conducted. First, we looked at item agreement. Teachers and observers had exact-match agreement on 55% of the log items, and exact-match or comparable agreement (defined as not today/little OR moderate/considerable) for 74% of the log items. Second, we conducted a correlational analysis to examine the relationship between teachers' and observers' responses on a given scale; this analysis indicated that teachers and observers report the most qualitatively similar student behaviors for *Representations* and *Connections*. These findings comparing the teachers' and observers' scores bring to light the issue of the perspective of the person completing the log (teacher versus observer) and the need to make these comparisons for a larger sample of lessons to adequately evaluate this assumption (Walkowiak et al., 2018).

Assumptions MS.S.2 and IM.S.2

We assume that scores resulting from each instrument represent the construct it is intended to measure. This assumption can be evaluated with evidence from the *Standards* based on test content and internal structure.

MS.S.2

In order to evaluate evidence based on content of the M-Scan, we solicited feedback from experts as to whether its content reflected standards-based mathematics teaching practices. The experts considered the suitability of each dimension, the relationships among the dimensions, and the appropriateness of the coding guide descriptors for each indicator (Walkowiak et al., 2014). Another source of evidence is the M-Scan's internal structure. During the development process, we theoretically mapped the dimensions onto four domains, as shown in Table 5.1. Confirmatory factor analyses were utilized (863 lessons in 281 classrooms of grades 3–5 teachers with varying experience levels AND 343 lessons in 139 classrooms of grades K-5 second-year teachers) to empirically test the theoretical structure. Results indicated that this theoretical structure is supported empirically as suggested by fit statistics, statistically significant factor variances, and statistically significant factor loadings (Walkowiak, 2018). This type of empirical evidence supports that the relationships amongst items align with the instrument's theoretical framework.

IM.S.2

Like the M-Scan, we solicited feedback from experts on the content of the IPL-M as to whether the items mapped to their assigned scale (e.g., *Math Talk*, *Problem Solving*). The experts considered the suitability of each item and the theoretical relationships among the items (Walkowiak et al., 2018). We also examined the internal structure of the IPL-M, expecting the empirical results to match the theoretical model with items mapping to the assigned scale (e.g., *Problem Solving*, *Math Talk*). Because of the nested nature of the data (lessons nested within teachers), we used a two-level confirmatory factor analysis (CFA) to test the theoretical structure of the IPL-M. In our sample, 139 teachers logged 5,170 lessons, with a mean of 37 lessons per teacher. The two-level CFA provided evidence regarding the empirical strength of the items and scales in explaining variance between and within teachers. Fit statistics (i.e., RMSEA, SRMR) suggested the data fit the theoretical model. Additionally, all item loadings were significant, ranging from .35 to .91, suggesting that the items each uniquely contribute to understanding differences between teachers on the given scale. In addition, we calculated Cronbach's alpha coefficients for each scale (Geldhof, Preacher, & Zyphur, 2014). These coefficients represent the extent to which the items on a scale measure a unitary construct (e.g., *Math Talk*). The coefficients, ranging from .80 to .93, supported the strength of the scales in representing students' opportunities to engage in behaviors aligned with the given construct (Walkowiak et al., 2018).

Assumptions MS.S.3 and IM.S.3

The third assumption under the scoring inference is that the scores represent variation in teaching practices. For both instruments, this assumption is supported by evidence based on response processes and internal structure.

MS.S.3

This assumption focuses on the appropriateness of the scoring; that is, M-Scan developers and users expect variation among teachers in their teaching practice, which should translate to coders using the full range of possible scores (1–7). To test this assumption, we examined evidence of quantitative response processes, including descriptive statistics such as standard deviations. The frequencies for most dimensions indicate less use of the "7" on the scale, but one would expect optimal demonstration to appear less often. For the experienced teacher sample, standard deviations across the dimensions ranged from 1.10–1.58 while the beginning teacher sample ranged from 1.07–1.90, suggesting adequate variation in scores for each sample. Another way to consider variation is to examine results of the CFA when we tested the internal structure. The CFA analysis showed the factor variances were statistically significant, suggesting that these empirically supported theoretical domains, and thereby M-Scan scores, explain variation between teachers in their instructional practices (Walkowiak, 2018).

IM.S.3

Similar to the evidence described for the M-Scan, we examined quantitative evidence of response processes, comparing the responses of 139 second-year teachers who completed 5,170 logs and 59 experienced teachers (five or more years) who completed 771 logs (Walkowiak et al., 2018). First, we expected teachers to use the full range of scores on the scale for the items. Frequency distributions indicated the teachers used the full range of scores; as expected, the proportions for "considerable" were much less than the proportions for "not today." Second, we examined the standard deviation for each item on the IPL-M, which showed substantial variation (median = .84, minimum = .48, maximum = 1.10), indicating heterogeneity in teacher responses for the log items. Finally, we examined the proportion of variance that lies between teachers on the log items by examining the item intraclass correlations (ICC). We used .10 as the minimum cutoff for substantive variation because this is common cutoff for variation to warrant the use of multilevel modeling (Raudenbush & Bryk, 2002). Results indicated that log items distinguish between teachers because substantial variation in item responses lies between teachers (second-year teachers: minimum ICC = .20, median = .29; experienced teachers: minimum ICC = .15, median = .38). These results suggest that the log items appear to discriminate between teachers (Walkowiak et al., 2018). Variation between teachers on the IPL-M is further supported by evidence based on internal structure. In the aforementioned CFA, all factor variances were significant, indicating the scales explain variation in mathematics instruction between teachers (Walkowiak et al., 2018).

Assumptions MS.S.4 and IM.S.4

The fourth assumption for the scoring inference relates to how the empirical data supports the theoretical structure. We use evidence based on internal structure for both measures as we did for the second and third assumptions.

MS.S.4

We hypothesized that bivariate correlations between the M-Scan dimensions would indicate relationships in theoretically supported ways, some at a higher strength than others. While not reported in any of our research reports, we examined bivariate correlations for the experienced and novice teacher data sets to test this hypothesis. For example, we expected a positive correlation between cognitive demand and the use of multiple representations. This hypothesis was confirmed for the experienced (.62) and beginning teacher data sets (.54). Second, we tested this assumption relative to the domains when conducting the CFAs. Factor (or domain) correlations were estimated as part of the CFAs; they were .70 or below, with two exceptions, suggesting that the domains represent different aspects of math instruction, which supports our theoretical structure. In both the experienced and novice teacher data sets, the correlations between tasks and discourse were 0.86 and 0.83, respectively. These correlations make sense theoretically in that teachers who use higher-quality tasks also engage their students in higher-quality discourse. Second, in the experienced teacher data set, the correlation between tasks and representations was 0.83, whereas the correlation for the novice teacher data set was 0.70. Similarly, these correlations are not surprising when one would expect teachers who use more high-quality tasks to also promote the use of multiple mathematical representations.

IM.S.4

Similar to the M-Scan, correlations across dimensions were examined within a two-level CFA for the IPL-M. All correlations were positive, and all correlations except one were below .72; we expected positive correlations because the dimensions of instruction are related. The correlation between *Math Talk* and *Problem Solving* was .93; this strong correlation is theoretically supported because one expects a teacher who gives more opportunities for math talk to also provide more opportunities for problem solving.

Assumption MS.S.5

The final assumption of the scoring inference only applies to the M-Scan and is one we have not tested, that scoring is bias-free. The assumption examines the extent to which systematic error due to various users' perspectives is minimized. Since a teacher completes the IPL-M for her own instruction, not someone else's, bias due to a teacher's perspective is not as much of a threat to validity as external coders' biases due to different experiences and backgrounds.

MS.S.5

We describe one form of bias that is particularly relevant to M-Scan coding: the expertise of the coder. Although coders participate in training to reach consistent understandings about applying the coding rubrics, there is potential bias related to expertise in standards-based mathematics teaching practices. For example, a doctoral student in mathematics education may have a deeper expertise in the theoretical underpinnings of standards-based mathematics instruction than an undergraduate student. As such, their observational scores on the same lesson may differ. To test the extent to which this type of undesirable variation exists, a generalizability study, where coders are purposefully selected to represent the varying levels of expertise prior to training, would indicate how much score variance is caused by the coders as related to level of expertise.

Comparison Summary: Scoring

For both measures, the scoring inference relates to the important assumption that scores are assigned accurately by the coder or teacher and that the content of the measure is supported empirically and theoretically. One critical difference relates to perspective. Teachers and coders have different perspectives. For the M-Scan, both the master and other coders have the perspective of an external observer in the classroom. However, for the IPL-M, an observer in the classroom and the teacher do not have the same perspective; the teacher is facilitating the lesson that is being logged while the observer only watches. It is important to consider this issue of "perspective" when evaluating the observer versus logging teacher data for the same lesson.

The Generalization Inference

The generalization inference examines if the number of sampled lessons is a reliable measure of instructional practices, meaning it represents a fair snapshot of the teacher's instruction as a whole. As recommended by Hill, Charalambous, and Kraft (2012), applying generalizability theory is important in understanding sources of score variation and in making scoring decisions such as how many lessons are needed to observe or log and how many coders are needed per lesson (for an observational protocol). The goal is to remove extraneous variation that is explained by factors other than the construct of measurement. When measuring teaching practice, there are a host of factors that explain score variation that one may want to consider in a generalizability study such as coder (for observational measures), duration of log completion (for teacher logs), day of week, and month of year. Table 5.4 presents two assumptions and accompanying validity evidence for the generalization inference for the M-Scan and IPL-M. We recommend a primer on generalizability theory by Shavelson and Webb (1991) for readers who want to deepen their understanding.

Assumptions MS.G.1, G.2 and IM.G.1, G.2

Assumptions G.1 and G.2 are paired together for each measure because both assumptions involve the application of generalizability theory. In the case of the M-Scan, we conducted one generalizability study (G-study) to determine the sources of score variation in an effort to minimize unexplained error (MS.G.2). This G-study has two related decision studies (D-study) to determine how many coders are needed per lesson (MS.G.1b) and how many lessons are needed per teacher to produce reliable scores (MS.G.1a). Similar to the M-Scan, we apply generalizability theory to examine how much variation is explained by various sources for each scale of the IPL-M to ensure that unexplained error is minimized (IM.G.2). We then determine how many days of logging are required to achieve a reliable estimate on all scales of the IPL-M (IM.G.1).

MS.G.1, MS.G.2

The G-study focuses on the sources of variance associated with the M-Scan dimension scores (i.e., classroom, coder, rubric, and interactions). We estimated the variance components across 60 mathematics lessons with two designs. The first design comprised 50 lessons, each coded by a single coder. The second design included 10 additional lessons, each coded by three independent coders. Results of the G-study indicated that the largest source of variance was the classroom (33%), the desired source. Coders (1%) and the classroom by coder interaction (5%) accounted for little variance (Walkowiak et al., 2014).

Table 5.4 Assumptions and Validity Evidence for the Generalization and Extrapolation Inferences for M-Scan and IPL-M

Assumptions: Generalization (G)	Validity Evidence
M-Scan (MS)	Reliability
MS.G.1*: The M-Scan can be reliably used to measure standards-based mathematics teaching practices based on video recordings of K-5 mathematics lessons. (a) The number of observed lessons is adequate to make generalizations about a teacher's practice. (b) The number of coders per lesson is adequate to make generalizations about coders' scores reliably representing what occurred in the lesson. MS.G.2: Unexplained error is minimized.	MS.G.1a: Decision study to determine the number of lessons to be observed to obtain a reliable estimate about the extent to which a teacher implements standards-based mathematics instruction. MS.G.1b: Decision study to determine the number of coders needed per lesson so that scores reliably represent standards-based mathematics teaching practices. MS.G.2: Generalizability study to determine the sources of variance (e.g., teacher, coder, rubric) and evaluate if we are sufficiently accounting for unexpected error.
IPL-M (IM)	Reliability
IM.G.1**: The scores on the IPL-M scales reliably measure mathematics instructional practices based on teachers' self-reported daily logs of the frequency and duration of students' opportunities to engage in particular behaviors. IM.G.2: Unexplained error is minimized.	IM.G.1: Decision study to determine the number of days of logging that are necessary to obtain a reliable estimate of a teacher's practice on dimensions measured by the IPL-M scales. IM.G.2: Generalizability study to determine variance explained by logging variables (e.g., item, teacher).

Assumptions: Extrapolation (E)	Validity Evidence	
M-Scan (MS)	Response Processes	Relations to Other Variables
MS.E.1: Scores on the M-Scan dimensions demonstrate theoretically supported correlations with other measures of instructional practice, teacher pedagogical content knowledge, and mathematics teaching efficacy.	MS.E.1: Interviews with students, parents, and principal support that the M-Scan scores align with perceptions of the teachers' instructional practices.	MS.E.1: Examination of the relationship between M-Scan scores and scores on: other measures of instructional practice, measures of teacher knowledge, and measures of teacher efficacy beliefs.

(*Continued*)

Table 5.4 (Continued)

M-Scan (MS)	Response Processes	Relations to Other Variables
MS.E.2: M-Scan scores are predictive of student achievement, student engagement, and teacher evaluation ratings.		MS.E.2: Examination of the relationship between M-Scan scores and measures of student achievement, teacher value-added scores, measures of student engagement, and teacher evaluation ratings (e.g., principal appraisals).

IPL-M (IM)	Response Processes	Relations to Other Variables
IM.E.1: Scores on the IPL-M demonstrate theoretically supported correlations with other measures of instructional practice, teacher pedagogical content knowledge, and mathematics teaching efficacy. IM.E.2: Scores on the IPL-M are predictive of student achievement and teacher evaluation scores.	IM.E.1: Interviews with students, parents, and principal support that the IPL-M scores align with perceptions of both teachers' instructional practices and students' learning opportunities.	IM.E.1: Examination of the relationship between IPL-M scores and scores on: other measures of instructional practice, measures of teacher knowledge, and measures of teacher efficacy beliefs. IM.E.2: Examination of the relationship between IPL-M scores and measures of student achievement, teacher value-added scores, and teacher evaluation ratings (e.g., principal appraisals).

*MS.G.1 stands for M-Scan, Generalization Inference, Assumption 1; subsequent labels have same pattern.
**IM.G.1 stands for IPL-M, Generalization Inference, Assumption 1; subsequent labels have same pattern.

We conducted one of the aforementioned associated D-studies and examined the number of coders needed per lesson. First, for the single-coder design, we calculated the generalizability coefficient (.94); however, this is a liberal appraisal of reliability because coder and coder inter-actions were not in the design as sources of variance due to having a single coder. Therefore, with the 10 additional lessons with three independent coders, we calculated the generalizability coefficient (.84) for the observed design and estimated the coefficient for one rater per lesson (.74). The results of this D-study suggest that one coder per lesson is sufficient and that coders score the rubrics consistently (additional support for assumption MS.S.1). The second D-study has not been conducted to determine the number of lessons needed for reliable scores representing a teacher's instructional practices.

IM.G.1, IM.G.2

For the IPL-M, we conducted a G-study to determine how much score variation is explained by various sources. Results indicated that the teacher, the desired source of variation, explained between 8% and 18% of the score variance depending upon the IPL-M scale. Forty-five percent to 55% of the score variance was attributed to random error; this finding is in line with other researchers who have found a substantial amount of "noise" related to measuring teaching practices (Hill et al., 2012).

The related D-study involved examining how many lessons are needed to obtain a reliable estimate of a teacher's practice on a given scale using the sample of 139 teachers who logged 5,170 lessons. Results indicate that the 10 days of logging results in a reliable estimate on the scales (generalizability coefficients > .75 except *Connections* = .66). Robustness checks were conducted using 22 days as the minimum (generalizability coefficients > .80 except *Connections* = .72) and established that 10 days of logging is sufficient.

Comparison Summary: Generalization

For the generalization inference, the application of generalizability theory is very relevant and particularly important (Hill et al., 2012). The primary distinction between applying generalizability theory to the M-Scan versus the IPL-M is differences in the sources of variance. In both cases, the variance source of interest is the teacher or classroom, and we included the rubric (M-Scan) or item (IPL-M) as a source of variance. Sources of variation that apply similarly to both M-Scan and IPL-M include: day of the week of the lesson and month in the academic year. Sources of variation that differ include: coder (M-Scan only), duration of the logging session (IPL-M only), and length of the video-recording (M-Scan only).

The Extrapolation Inference

The extrapolation inference makes claims about the extent to which scores are related to a broader conception of instructional practice. In other words, we generally seek to understand how scores on the M-Scan and IPL-M are related to other indicators of teacher expertise that have been shown to be related to instructional practices (e.g., teacher knowledge, teaching efficacy). Table 5.4 presents the sets of assumptions (two per measure) that accompany the extrapolation inference. The types of validity evidence that support these assumptions are response processes and relations to other variables for both measures.

Assumptions MS.E.1 and IM.E.1

We expect that scores on measures of instructional practices correlate positively with other measures of instructional practice, which is examined using two types of evidence: response processes and relations to other variables. In interviews with students, parents, and a principal about their perceptions of a teacher's instructional practices, we expect alignment (or not in the case of parents) with M-Scan and IPL-M scores, particularly if we asked specific questions. For example, if students were asked to describe their opportunities to talk about mathematics, we may see some patterns between M-Scan and IPL-M scores and students' descriptions. On the other hand, interviews to include parents' perceptions may not align with all dimensions on the measures. Parents are not in the classroom every day; therefore, their perceptions of day-to-day practices may be influenced by other variables (e.g., teacher communication).

The second type of evidence is related to other variables. For both measures, we expect divergent and convergent evidence with other measures of teaching performance. We also expect to see positive correlations to PCK and efficacy beliefs; one exception may be the *Mathematical Procedures* scale on the IPL-M. We have investigated M-Scan's relationship to other teaching measures. For a set of 60 lessons, we calculated bivariate correlations to examine the relationship between M-Scan, CLASS (Classroom Assessment Scoring System; Pianta et al., 2008), a measure of emotional support, classroom organization, and instructional support, and RTOP (Reformed Teaching Observation Protocol; Piburn et al., 2000), a measure of inquiry-based instruction. As hypothesized there were higher correlations between RTOP and M-Scan dimensions than between CLASS domains and M-Scan dimensions (Walkowiak et al., 2014).

Assumptions MS.E.2 and IM.E.2

For the M-Scan, we expect scores to be predictive of student achievement and teacher evaluation ratings (depending upon the type(s) of student achievement test and of the instructional components on the teacher evaluation rating rubric). We have investigated the relationship between M-Scan and student achievement (Ottmar, Rimm-Kaufman, Larsen, & Berry, 2015). M-Scan scores were positively related to third-grade state standardized assessment, but only for the intervention group whose teachers had received training in a socioemotional intervention. Further investigation is warranted to better understand the relationship using other measures of student performance. Similarly, the same investigations should be conducted with the IPL-M to support this assumption.

Comparison Summary: Extrapolation

The extrapolation assumptions and the types of evidence for both instruments are very similar; concurrent and predictive relationships are central. Overall, the hypothesized correlations with other measures are similar across the measures. However, two important differences in the measures exist, both previously discussed. The perspective is different in that the teacher completes the log whereas a rater completes the observation, and the focus is different in that the M-Scan focuses on quality of teaching practices whereas the IPL-M focuses on frequency and duration. It would be interesting to unpack the extent to which, if at all, these differences influence correlations with other variables. With the exception of the M-Scan's relationship to other measures of teaching practice, we have not engaged in these analyses. However, these analyses are warranted given the importance of understanding these relationships.

The Implications Inference

The implications inference evaluates the fit between the score interpretations and the implications associated with those scores. Implications can be conceptualized as: (1) users' interpretations about the trait being measured (i.e., standards-based mathematics instruction) and (2) the ways in which the scores are used, which is why the implications inference is also referenced as the score use inference. To ensure that users' interpretations based on scores are appropriate, it is critical to ensure that scores represent meaningful variation in the trait being measured (Kane, 2006). Furthermore, to evaluate the appropriateness of the ways in which scores are used, the actual uses of the scores should align with the intended uses. While it is not possible for measure developers to prevent misuse in all situations, developers should explicitly state the intended interpretations and uses to minimize misuse. This can be supported through effectively designed score reporting platforms and availability of interpretive guides (O'Leary et al., 2017). For the M-Scan and IPL-M, we identify three types of validity evidence, including evidence based on response processes, relations to other variables, and score consequences. The assumptions for the implications inference, displayed in Table 5.5, are similar for the two measures with the exception of the third assumption (IM.I.3) for the IPL-M.

Assumptions MS.I.1 and IM.I.1

The assumption that the scores support meaningful judgments about teaching performance is a central tenet of the validity argument presented in this chapter. However, we did not collect evidence for either instrument that explicitly connects to this assumption. Evidence that supports the implication inference includes asking school leaders or instructional coaches if the observed scores on the M-Scan or IPL-M align with their perceptions of a teacher's practice. One way to generate this type of evidence is to qualitatively interview instructional leaders

Table 5.5 Assumptions and Validity Evidence for the Implications Inference for M-Scan and IPL-M

Assumptions: Implications (I)	Validity Evidence		
M-Scan (MS)	Response Process	Relations to other variables	Consequences
MS.I.1*: The observed scores on the lesson support the implications associated with the judgments about teaching performance (Bell et al., 2012). MS.I.2: The observed scores help inform improvements to programs (e.g., professional development, teacher preparation) and teachers' instructional practices.	MS.I.2a: The ways in which scores are reported support users in making appropriate interpretations and using the scores appropriately.	MS.I.1: The observed M-Scan scores correlate positively with other measures of teaching performance (e.g., principal evaluation, IPL-M)	MS.I.2b: Users utilize M-Scan scores to identify areas for program redesign and instructional improvement.
IPL-M (IM)	Response Process	Relations to other variables	Consequences
IM.I.1:** The observed scores across lessons support the implications associated with the judgments about teaching performance (Bell et al., 2012). IM.I.2: The observed scores help inform improvements to programs (e.g., professional development, teacher preparation) and teachers' instructional practices. IM.I.3: Completing the IPL-M encourages teachers to reflect on their practice.	IM.I.2a: The ways in which scores are reported support users in making appropriate interpretations and using the scores appropriately.	IM.I.1: The observed IPL-M scores correlate positively with other measures of teaching performance (e.g., principal evaluation, M-Scan)	IM.I.2b: Users utilize IPL-M scores to identify areas for program redesign and instructional improvement. IM.U.3: Teachers report that completing the IPL-M provides them an opportunity to reflect on their instruction, but does not necessarily change their instruction in systematically desirable or undesirable ways.

*MS.I.1 stands for M-Scan, Implications Inference, Assumption #1; subsequent labels have same pattern.
**IM.I.1 stands for IPL-M, Implications Inference, Assumption #1; subsequent labels have same pattern.

and rank teachers based on the leaders' perceptions. Another way is to show leaders videos of teachers' instruction and ask them to rank those teachers based on their alignment to standards-based instruction. In either case, a comparison of the resulting ranking and the ranking based on the M-Scan or IPL-M scores provides evidence to support that the implications are valid. These comparisons, which are similar to an analysis of value-added measures and principal evaluations conducted by Grissom, Loeb, and Doss (2016), reveal the extent to which the M-Scan and/or IPL-M scores reflect meaningful variation in practice in ways that support the implications associated with the judgments about teaching performance (Bell et al., 2012).

Assumptions MS.I.2 and IM.I.2

A main implication of the M-Scan and IPL-M scores is that the scores are utilized to inform improvements to programs and teachers' instructional practice. The scores are not intended to be used for high-stakes decisions such as teacher retention or merit pay. Therefore, it is critical to ensure that the scores are interpreted and used appropriately (MS.I.2a and IM.I.2a), and inform program redesign and instructional improvement (MS.I.2b and IM.I.2b). One way to collect this type of evidence, ensuring that users interpret and use scores appropriately, is based on response processes. For example, it is important to interview or survey users about how they interpret the scores presented on a scoring report or another data output format. In addition, evidence of response process (i.e., interviews) supports how those scores will be used and if those uses are appropriate.

Developers collect evidence to understand the extent to which feedback given directly to teachers or to another user informs effective changes to program redesign or instructional practice. For example, if we hope that providing teachers with their scores on M-Scan or IPL-M informs changes in their practice, then we need to document those changes by collecting evidence of instructional practice. One way to collect this evidence is by examining M-Scan or IPL-M scores across time in a randomized experiment where: a comparison group of teachers receives their M-Scan or IPL-M scores along with coaching (Group A); and a control group does not receive M-Scan or IPL-M scores or coaching (Group B). It is hypothesized that Group A would improve their instruction at a higher rate than Group B. This evidence supports that the M-Scan or IPL-M scores are interpreted in ways that have meaningful consequences for instruction.

Assumption IM.I.3

Completing the IPL-M every day for 10 or more days likely encourages teachers to reflect on their practice. When we conducted cognitive interviews for the IPL-M, a few teachers informally commented on liking the opportunity to reflect on their instruction. However, without data to understand the trends in their instruction (e.g., little amounts of math talk), it is very unlikely that the brief reflection changes their instruction in systematic ways. To test this claim, a more in-depth study with a larger number of logging teachers is necessary. A need exists for a study focused on how teachers perceive the log as a reflective tool along with observations in their classroom to detect if there are any systematic changes in their instruction due to reflection through the IPL-M.

Comparison Summary: Implications

There are two key distinctions between the M-Scan and IPL-M when we consider the implications inference. First, when comparing judgments by principals to scores on the M-Scan versus the IPL-M, it is important to remember the distinction between the two instruments. The M-Scan measures *quality* while the IPL-M measures *frequency*. Consequently, when comparing observation or log scores to principal ratings, it is important to ask the principal to examine teaching performance from different angles. For example, when comparing mathematical discourse community on M-Scan to a principal's rating of the discourse, it would be important that the principal consider both the presence of the discourse and the quality. In contrast, for the IPL-M, a principal would want to focus on the frequency of talk in the classroom *about mathematics*. Second, similar considerations should be considered to inform instructional improvements. As data are used to inform instruction or program improvements, it is extremely important that teachers or program evaluators are interpreting the scores accurately based upon what the scores are measuring (quality versus frequency and duration).

Discussion

We have presented interpretation-use arguments (IUAs) for two different instruments that measure mathematics teaching practices, the M-Scan, an observational protocol, and the IPL-M, an instructional log. We used Kane's (2013, 2006) framework by presenting the key inferences (scoring, generalization, extrapolation, and implications) and supporting assumptions for each instrument. We described the validity argument by providing the types of evidence from the *Standards* (AERA et al., 2014) for each of the assumptions. We highlight four key contributions of this work in our discussion, pointing to lessons we have learned by engaging in validation work.

The first significant outcome of this work is pairing Kane's framework with the *Standards* to build our arguments. Kane's explicit focus on the interpretations of scores is an important reminder for those developing an instrument. We argue that the intended interpretations and uses of the scores is neglected too often in validation work. As previously noted, many existing studies use the *Standards*' evidence types to organize validity arguments. This approach focuses attention on each evidence type rather than on how the multiple sources of evidence are collectively linked to the interpretations of data through an evidence-centered design (Mislevy & Haertel, 2006). This lack of attention to score use and interpretation is dangerous because misinterpretation or misuse of scores could have negative implications for teachers and how their work is described and possibly evaluated. By overlaying Kane's interpretation-use approach with the evidence types in the *Standards*, the proposed interpretations and uses remain central, but the evidence types provide a tangible way to organize the argument. By providing the two example measures in this chapter and pointing out their distinctions, the field of mathematics education, and education in general, gain models for how validity arguments should be constructed, specifically in the context of measuring teaching practices.

A second key contribution of this chapter is our focus on the measurement of teaching practices, a different endeavor and less straightforward than the measurement of constructs like knowledge and beliefs. Capturing, analyzing, and describing mathematics instruction is critical for understanding the types of learning opportunities provided to students. Past research has indicated a wide degree of variation in the nature and quality of mathematical learning opportunities provided to students (Hiebert et al., 2005). Identifying and understanding this variation has implications for allocating resources and focusing instructional improvements to ensure high-quality mathematical learning opportunities for all students (Boston & Wilhelm, 2017). However, without data that produces valid inferences, we cannot identify and understand this variation, pointing to the importance of the examples presented in this chapter.

Beyond the need to understand variation, there has been recent heightened interest in the measurement of teaching practices due to its inclusion in teacher evaluation systems (Danielson, 2013). The measurement of teaching practices for the purposes of personnel evaluation and decisions is risky in situations where the validity of data, its interpretation, and how it is used have not been carefully laid out. This chapter addresses how instrument developers for such purposes might consider articulating and evaluating their validity arguments.

Articulating a coherent argument for validity is critically important in instrument development, bringing us to the third contribution of this chapter. The biggest lesson we learned in our development processes is to be *intentional* with the collection of validity evidence from the start by laying out the interpretation-use argument up front. As presented earlier, Table 5.2 shows the types of evidence for each inference for the M-Scan and IPL-M. How one chooses to measure teaching practices (e.g., observations, teacher-reported data) has implications for the validity arguments that are built and how they are tested. In the case of the M-Scan and IPL-M, while similar in their interpretation-use arguments, there are clear distinctions. For example, there is an issue of perspective. The IPL-M is completed by the teacher, the person who is implementing the instruction while the M-Scan is coded by an observer who has a different perspective as someone viewing the classroom activities, not implementing them.

As stated, we have not collected evidence for every assumption that we outlined. We made decisions about collecting evidence based upon the assumptions that were the most important (e.g., accurate scoring) in our interpretation-use arguments. However, this does not preclude us or other researchers from collecting additional evidence of validity for the M-Scan and IPL-M (e.g., examining relationships to student achievement outcomes). We encourage other developers of measures of teaching practice to be especially careful as they prioritize assumptions and inferences to be evaluated, keeping the intended interpretation and use of the scores central and explicit.

The fourth and final contribution of this chapter focuses on the reality of challenges when conducting validation studies. For both the M-Scan and IPL-M, we were both rigorous and realistic in our collection of validity evidence, but sometimes, these goals conflicted. There are time, fiscal, and human demands when executing the collection and evaluation of the evidence described in this chapter. Related to these demands, access to a large number of participants is necessary to generate the amount of data necessary to collect some of the evidence types (e.g., G- and D-study data). Furthermore, there sometimes exists a tension between moving research studies along at an appropriate pace to generate knowledge for dissemination and the need to conduct rigorous validation studies. Sometimes, these challenges prevent instrument developers from collecting certain types of evidence, again pointing to the need to prioritize assumptions to be evaluated based upon the intended interpretations and uses of the resulting data. We also emphasize that validation studies are iterative and ongoing (Beckman, Mandrekar, Engstler, & Ficalora, 2009). Developers should continuously revisit their interpretation-use argument and continue to collect validity evidence as they move forward in their work, and perhaps most importantly, developers must clearly explicate how and by whom their measure is intended to be used. As described earlier, these instruments were designed to be used in elementary (K-5) classroom settings; future work should address the validity of the data from these instruments in secondary classroom settings.

Our focus in this chapter on the M-Scan and IPL-M makes validation studies for measures of teaching practice more straightforward and highlights differences in validity arguments for the two types of instruments. We provide a pragmatic approach to articulating validity arguments. By pairing Kane's framework with the evidence types from the *Standards*, we offer comprehensive, systematic examples for the field of educational research.

Acknowledgments

This work is funded by the National Science Foundation under Award #1118894 and by the Institute of Education Sciences, U.S. Department of Education under Grant R305A070063. Any opinions, findings, and conclusions or recommendations expressed in this material are those of the authors and do not necessarily reflect the views of NSF or the U.S. Department of Education.

References

Adams, E. L., Carrier, S. J., Minogue, J., Porter, S. R., McEachin, A., Walkowiak, T. A., & Zulli, R. A. (2017). The development and validation of the Instructional Practices Log in Science: A measure of K-5 science instruction. *International Journal of Science Education*, 39(3), 335–357. doi:10.1080/09500693. 2017.1282183

American Educational Research Association, American Psychological Association, & National Council on Measurement in Education. (2014). *Standards for educational and psychological testing*. Washington, DC: AERA.

Beckman, T. J., Mandrekar, J. N., Engstler, G. J., & Ficalora, R. D. (2009). Determining reliability of clinical assessment scores in real time. *Teaching and Learning in Medicine*, 21(3), 188–194. doi:10.1080/10401330903014137

Bell, C. A., Gitomer, D. H., McCaffrey, D. F., Hamre, B. K., Pianta, R. C., & Qi, Y. (2012). An argument approach to observation protocol validity. *Educational Assessment*, 17(2–3), 62–87. doi:10.1080/10627197.2012.715014

Berry, R. Q., Rimm-Kaufman, S. E., Ottmar, E. M., Walkowiak, T. A., Merritt, E., & Pinter, H. H. (2013). *The Mathematics Scan (M-Scan): A measure of standards-based mathematics teaching practices* (Unpublished measure). Charlottesville, VA: University of Virginia.

Borko, H., Stecher, B. M., Alonzo, A., Moncure, S., & McClam, S. (2005). Artifact packages for characterizing classroom practice: A pilot study. *Educational Assessment*, 10(2), 73–104. doi:10.1207/s15326977ea1002_1

Boston, M. D., & Wilhelm, A. G. (2017). Middle school mathematics instruction in instructionally focused urban districts. *Urban Education*, 52(7), 829–861. doi:10.1177/0042085915574528

Camburn, E., & Barnes, C. A. (2004). Assessing the validity of a language arts instruction log through triangulation. *The Elementary School Journal*, 105(1), 49–73. doi:10.1086/428802

Cohen, J., & Goldhaber, D. (2016). Building a more complete understanding of teacher evaluation using classroom observations. *Educational Researcher*, 45(6), 378–387. doi:10.3102/0013189X16659442

Danielson, C. (2013). *The framework for teaching evaluation instrument: 2013 edition*. Princeton, NJ: The Danielson Group. Retrieved from www.danielson group.org/framework/

Elliott, S. N. (2015). Measuring opportunity to learn and achievement growth: Key research issues with implications for the effective education of all students. *Remedial and Special Education, 36*(1), 58–64. doi:10.1177/0741932514551282

Geldhof, G. J., Preacher, K. J., & Zyphur, M. J. (2014). Reliability estimation in a multilevel confirmatory factor analysis. *Psychological Methods, 19*(1), 72–91.

Grissom, J. A., Loeb, S., & Doss, C. (2016). The multiple dimensions of teacher quality: Does value-added capture teachers' nonachievement contributions to their schools? In J. A. Grissom & P. Youngs (Eds.), *Improving teacher evaluation systems: Making the most of multiple measures* (pp. 37–50). New York, NY: Teacher College Press.

Hayes, K. N., Lee, C. S., DiStefano, R., O'Connor, D., & Seitz, J. (2016). Measuring science instructional practice: A survey tool for the age of NGSS. *Journal of Science Teacher Education, 27,* 137–164. doi:10.1007/s10972-016-9448-5

Hiebert, J., Stigler, J. W., Jacobs, J. K., Givvin, K. B., Garnier, H., Smith, M., . . . Gallimore, R. (2005). Mathematics teaching in the United States today (and tomorrow): Results from the TIMSS 1999 video study. *Educational Evaluation and Policy Analysis, 27*(2), 111–132. doi:10.3102/01623737027002111

Hill, H. C., Blunk, M., Charalambous, C., Lewis, J., Phelps, G. C., Sleep, L., & Ball, D. L. (2008). Mathematical knowledge for teaching and mathematical quality of instruction: An exploratory study. *Cognition and Quality of Instruction, 26*(4), 430–511. doi:10.1080/07370000802177235

Hill, H. C., Charalambous, C. Y., & Kraft, M. A. (2012). When rater reliability is not enough: Teacher observation systems and a case for the generalizability study. *Educational Researcher, 41*(2), 56–64. doi:10.3102/0013189X12437203

Jacobs, V. R., & Spangler, D. A. (2017). Research on core practices in K-12 mathematics teaching. In J. Cai (Ed.), *Compendium for research in mathematics education* (pp. 824–852). Reston, VA: National Council of Teachers of Mathematics.

Kane, M. T. (2006). Validation. In R. L. Brennan (Ed.), *Educational measurement* (pp. 17–64). Westport, CT: American Council on Education and Prager Publishers.

Kane, M. T. (2013). The argument-based approach to validation. *School Psychology Review, 42*(4), 448–457. doi:10.1037/0033-2909.112.3.527

Kurz, A., Elliott, S. N., Kettler, R. J., & Yel, N. (2014). Assessing students' opportunity to learn the intended curriculum using an online teacher log: Initial validity evidence. *Educational Assessment, 19*(3), 159–184. doi:10.1080/106 27197.2014.934606

Messick, S. (1995). Validity of psychological assessment: Validation of inferences from persons' responses and performances as scientific inquiry into score meaning. *American Psychologist, 50*(9), 741–749.

Meyer, J. P., Cash, A. H., & Mashburn, A. (2011). Occasions and the reliability of classroom observations: Alternative conceptualizations and methods of analysis. *Educational Assessment, 16*(4), 227–243. doi:10.1080/10627197.2011.638884

Mislevy, R. J., & Haertel, G. D. (2006). Implications of evidence-centered design for educational testing. *Educational Measurement: Issues and Practice, 25*(4), 6–20. doi:10.1111/j.1745-3992.2006.00075.x

National Council of Teachers of Mathematics. (2000). *Principles and standards for school mathematics.* Reston, VA: NCTM.

National Governors Association Center for Best Practices, & Council of Chief State School Officers. (2010). *Common core state standards for mathematics.* Washington, DC: Common Core State Standards Initiative.

O'Leary, T. M., Hattie, J. A., & Griffin, P. (2017). Actual interpretations and use of scores as aspects of validity. *Educational Measurement: Issues and Practice*, *36*(2), 16–23. doi: 10.1111/emip.12141

Ottmar, E. R., Rimm-Kaufman, S. E., Larsen, R. A., & Berry, R. Q. (2015). Mathematical knowledge for teaching, standards-based mathematics teaching practices, and student achievement in the context of the Responsive Classroom approach. *American Educational Research Journal*, *52*(4), 787–821.

Pianta, R. C., La Paro, K., & Hamre, B. K. (2008). *Classroom Assessment Scoring System (CLASS)*. Baltimore, MD: Paul H. Brookes.

Piburn, M., Sawada, D., Turley, J., Falconer, K., Benford, R., Bloom, I., & Judson, E. (2000). *Reformed teaching observation protocol (RTOP): Reference manual* (ACEPT Technical Report No. IN00-3). Tempe, AZ: Arizona Collaborative for Excellence in the Preparation of Teachers.

Praetorius, A. K., Pauli, C., Reusser, K., Rakoczy, K., & Klieme, E. (2014). One lesson is all you need? Stability of instructional quality across lessons. *Learning and Instruction*, *31*, 2–12. doi:10.1016/j.learninstruc.2013.12.002

Raudenbush, S. W., & Bryk, A. S. (2002). *Hierarchical linear models: Applications and data analysis methods*. Thousand Oaks, CA: Sage Publications, Inc.

Rowan, B., & Correnti, R. (2009). Studying reading instruction with teacher logs: Lessons from the study of instructional improvement. *Educational Researcher*, *38*(2), 120–131. doi:10.3102/0013189X09332375

Rowan, B., Harrison, D., & Hayes, A. (2004). Using instructional logs to study mathematics curriculum and teaching in the early grades. *The Elementary School Journal*, *105*(1), 103–127. doi:10.1086/428812

Shavelson, R. J., & Webb, N. M. (1991). *Generalizability theory: A primer*. Thousand Oaks, CA: Sage.

Tourangeau, R., Rips, L. J., & Rasinski, K. (2000). *The psychology of survey response*. New York, NY: Cambridge University Press.

Walkington, C., Arora, P., Ihorn, S., Gordon, J., Walker, M., Abraham, L., & Marder, M. (2012). *Development of the UTeach observation protocol: A classroom observation instrument to evaluate mathematics and science teachers from the UTeach preparation program*. Unpublished manuscript, Southern Methodist University.

Walkowiak, T. A. (2018). *Utilizing confirmatory factor analysis to confirm the theoretical structure of measures of mathematics instruction: The case of the M-Scan*. Manuscript in preparation.

Walkowiak, T. A., Adams, E. L., Porter, S. R., Lee, C. W., & McEachin, A. (2018). *The development and validation of the Instructional Practices Log in Mathematics (IPL-M)*. Manuscript submitted for publication.

Walkowiak, T. A., Berry, R. Q., Meyer, J. P., Rimm-Kaufman, S. E., & Ottmar, E. R. (2014). Introducing an observational measure of standards-based mathematics teaching practices: Evidence of validity and score reliability. *Educational Studies in Mathematics*, *85*(1), 109–128. doi:10.1007/s10649-013-9499-x

Walkowiak, T. A., & Lee, C. W. (2013). *The Instructional Practices Log in Mathematics (IPL-M)*. Unpublished measure. Raleigh, NC: North Carolina State University.

Willis, G. B. (2005). *Cognitive interviewing: A tool for improving questionnaire design*. Washington, DC: Sage Publications, Inc.

6 Design and Validation Arguments for the Student Survey of Motivational Attitudes towards Statistics (S-SOMAS) Instrument

Douglas Whitaker, Alana Unfried, and Marjorie Bond

Attitudes toward statistics are important outcomes of statistics courses because they are linked with student achievement and because they affect students' lasting impressions of the discipline (Gal, Ginsburg, & Schau, 1997; Ramirez, Schau, & Emmioğlu, 2012). In short, "People forget what they do not use. But attitudes 'stick'" (Ramirez et al., 2012, p. 57). As the field of statistics education matures, instruments for measuring attitudes have been precipitated by an evolving understanding about the learning and teaching of statistics and the needs of researchers. The Surveys Of Motivational Attitudes towards Statistics (SOMAS) project is developing a family of instruments for assessing attitudes toward statistics for both of the aforementioned reasons: significant advances in the discipline have occurred since the release of the last major attitude assessment instrument in statistics education and current instruments are not designed for contemporary research needs.

This chapter outlines the development process of an instrument designed to be taken by students enrolled in undergraduate introductory-level statistics courses (S-SOMAS), including a brief overview of the motivation for developing a new instrument, the instrument development plan, and the validity evidence that will support the intended uses of the instrument. While the SOMAS project also includes instruments to be completed by instructors, only the S-SOMAS is described here. The SOMAS project is an ongoing instrument development project and specific details and goals may evolve or change as the project continues.

Validity evidence to be collected supporting the intended uses of the S-SOMAS instrument is organized using the sources of evidence detailed in the 2014 *Standards for Educational and Psychological Testing* document (American Educational Research Association, American Psychological Association, National Council on Measurement in Education [AERA, APA, & NCME], 2014). This chapter seeks to make explicit claims that may remain implicit in some validity studies. Together with a clearly articulated purpose for developing and using an instrument, these explicit claims and evidence statements represent a roadmap for the validation studies to be conducted for the S-SOMAS. By presenting the validity evidence claims and clear motivation for developing the S-SOMAS in one location, we hope to illuminate the process of gathering validity evidence during instrument development.

Motivation for Developing a New Instrument

Instruments for measuring attitudes toward statistics have been developed and used for decades. Today, the Survey of Attitudes Toward Statistics (SATS) instruments (Schau, 1992, 2003b) are among the most widely used and researched attitude surveys in the statistics education literature since their initial release and subsequent updates (Nolan, Beran, & Hecker, 2012). One factor that contributes to the popularity of the SATS is that it is claimed to be congruent with expectancy value theory (EVT) when few extant instruments in statistics education are aligned with an established educational theory (Ramirez et al., 2012; Schau, Millar, & Petocz, 2012). There are two versions of the SATS: a 28-item version that measures four constructs related to statistics attitudes (SATS-28; Schau, 1992) and a 36-item version that is the SATS-28 with eight additional items measuring two additional constructs (SATS-36; Schau, 2003b). The SATS-36 contains the SATS-28 as a subset of items. When statements apply to both instruments, the term SATS will be used.

The SATS have been widely used (e.g., Dauphinee, Schau, & Stevens, 1997; Schau & Emmioğlu, 2012; Vanhoof, Kuppens, Sotos, Verschaffel, & Onghena, 2011), and numerous limitations and challenges have been identified as they have been studied more. Notably, the alignment of the SATS instruments to EVT was post-hoc and did not guide the development of the instruments (Schau, 2003a), limiting the extent to which the instruments measure the theoretical constructs that comprise the EVT framework. The six constructs measured by the SATS-36 are Affect, Cognitive Competence, Value, Difficulty, Interest, and Effort (Schau, 2003b). Schau (2003b) noted that these constructs are not the same as the EVT constructs described in the EVT literature (e.g. Eccles, 1983, 2014). While the alignment of the SATS instruments to EVT was innovative in statistics education when they were developed, their widespread use has highlighted areas for future research and improvement.

Additionally, the SATS instruments were developed for use with students in introductory statistics classes at the undergraduate and graduate levels (Schau, Stevens, Dauphinee, & Vecchio, 1995), but contemporary research in statistics education has expanded to include research with many populations beyond undergraduate and graduate students, such as instructors of statistics, students in grades K-12 (though primarily focused on students in high school), teachers of students in grades K-12, preservice teachers, and learners of statistics in specialized contexts such as health sciences.

Adapting the SATS instruments so that they are suitable for all contexts is not a straightforward task: some items included on the SATS instruments are written in such a way that they presuppose that the respondent is enrolled in a statistics course. Additionally, two forms for each SATS instrument exist: a version intended to be administered near the beginning of the course, and a version intended to be administered near the end of the course. Adapting the instrument for use with groups beyond students enrolled in tertiary first courses in statistics would require at least three substantial types of changes. First, at least 25% of the items on the SATS-36 that explicitly position the respondent as being enrolled in a course; the items reference courses, tests, and problems and would need to be heavily modified or replaced. Second, other items implicitly position the respondent as being a student (e.g., by asking about future employability), which would require more than minor alterations. Third, the way some scales assess the underlying latent constructs would need to be reconceptualized (e.g., all four items in the Effort construct reference enrollment in some way). While some researchers have used the SATS instruments with other populations of interest, there is a lack of validity evidence for these uses because of the aforementioned challenges and concerns, and piecemeal modification of the instrument is problematic. Instead, new instruments are needed that have been designed for use with these populations.

Developing the S-SOMAS Instrument

The researchers working on the SOMAS project (SOMAS project team) all have prior experience researching attitudes in statistics education and using the SATS instruments. Before the decision to develop a new family of instruments was made, the team had considered working with the lead developer of the SATS instruments to update, modify, and adapt them to address concerns and for use with new populations. However, the team chose to develop new instruments because (1) using an existing educational framework to guide development will result in stronger instruments and (2) previous experiences using and researching the SATS instruments have suggested that there are other challenges to using these instruments beyond those already reported in the literature. The SOMAS project team instead chose to draw on their accumulated knowledge of the SATS to inform the development of new instruments.

The SOMAS project team is currently developing three online instruments: one intended to measure the attitudes toward statistics of undergraduate students enrolled in introductory statistics courses, the S-SOMAS, and two intended to be completed by instructors, thus enabling researchers to analyze attitudes in an introductory statistics course in a broader and more cohesive way than previously possible. While the project team anticipates that there will be a desire to use the instruments with other populations, restricting the focus to introductory college-level statistics will help ensure the development of high-quality instruments with strong validity evidence supporting their use with this population rather than attempting to meet all possible project goals simultaneously.

This chapter will focus on the development of the S-SOMAS instrument, intended to be used with undergraduate students. A brief overview of the development process—both planned and enacted—will be presented first. Then, the theoretical framework adopted for use with the SOMAS instruments, EVT, will be briefly described. Lastly, the plan for collecting validity evidence supporting the use and interpretations of the S-SOMAS will be presented.

Overview of Development Process

The development of the S-SOMAS instrument is a multiyear process that began May 2017 and is anticipated to conclude in 2020 (Figure 6.1).

There are numerous factors that contribute to the length of time needed for developing an instrument, including both the realities of professional life and the principles of instrument development. The development process illustrated in Figure 6.1 reflects an iterative development process wherein data collected lead to revisions of work done previously as well as a strong focus on collecting validity evidence to support the intended uses of the S-SOMAS instrument. Note that Figure 6.1 is meant to illustrate the sequencing of key events in the development process rather than a timeline to be interpreted on a linear scale. The primary activities in Figure 6.1 are the development of and revisions to the theoretical framework (described below), the development of and revisions to items, data collection using assembled forms, and analysis and interpretation of collected data to inform revisions. In addition to these primary activities, the collection of other data through focus groups and subject-matter expert (SME) review are included. Each of these development activities supports collection of validity evidence for the use and interpretations of the S-SOMAS instruments and represents an intentional development decision in the planning phase. The SOMAS project team anticipates having a finalized version of the S-SOMAS instrument prior to Fall 2020: this would represent the initial instrument development, two rounds of pilot data collection and revisions, and the operational administration to ensure the instrument is performing as intended with no subsequent changes planned. Note that a target deadline of Fall 2020 was not chosen first and the development process designed to meet this goal; rather, the development activities necessary to achieve an instrument with appropriate validity evidence for our intended purpose were determined and a timeline constructed around these activities. However, data collection

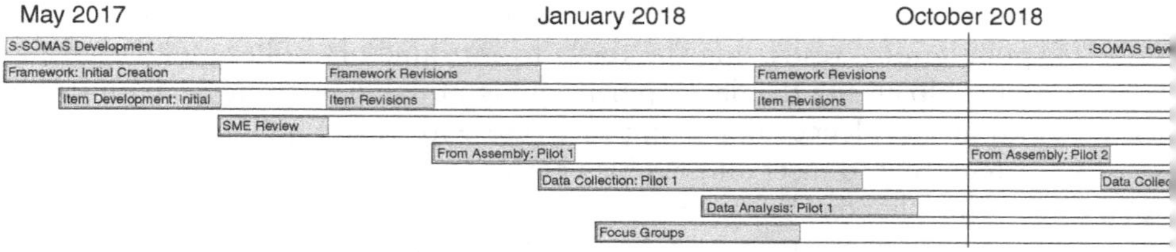

Figure 6.1 Timeline of S-SOMAS development.

with possible revisions in Fall 2020 is included in the Figure 6.1 timeline in recognition that more changes—even minor tweaks—may be necessary, and a large administration of the S-SOMAS is planned for Fall 2020.

While not intrinsic to the work of developing instruments, professional realities still affect the development timeframe and should be reasonably accounted for in planning. Examples of professional realities that may contribute to long development timelines include high teaching loads, time commitments for other projects, supplemental work such as writing proposals for grant applications, and the sequential development process introducing bottlenecks resulting from the division of labor based on team members' expertise.

Overview of the Theoretical Framework

The SOMAS project team has adopted EVT as the theoretical framework for the S-SOMAS instrument. EVT is a theory of motivation, and this theoretical focus is reflected in the term "motivational attitudes" in the SOMAS project name. Historically, researchers in statistics education have been interested in studying what has been referred to as *attitudes*, even when nominally aligned with established educational theories. The SOMAS project team uses the phrase *motivational attitudes* as a bridge to connect the field of statistics education's historical focus on attitudes to the broader educational literature of motivation that informs this project.

Before presenting the S-SOMAS EVT model, it is important to note that there are several related technical terms used throughout this chapter that are closely related: construct, scale, and factor. We conceive of *constructs* as a type of latent variable, that is, a random variable that cannot be measured directly (Raykov & Marcoulides, 2011). For each construct, a group of items—known as a *scale*—will ultimately be developed to measure it. While the term scale is sometimes conflated with the term instrument, we use the term scale to mean a group of items that assess a construct (Raykov & Marcoulides, 2011); the S-SOMAS instrument is then composed of several scales. When analyzing items and scales, the term *factor*

Framework Revisions	Framework Revisions (if needed)
Item Revisions	Item Revisions (if needed)
SME Review	
Form Assembly	Form Assembly: Operat
Data Collection: Operation	Data Collection: Operation
ta Analysis: Pilot 2	Data Analysis: Operation	Data Analysis: Operation
cus Groups (if needed)	

is used to denote an unobservable variable that explains the relationships among items (Raykov & Marcoulides, 2011). Ideally the unobservable variable represented by a factor would be a construct of interest, but this is not always the case.

The EVT framework explains achievement-related outcomes by relating them to an individual's beliefs about success on a given task (expectancies) and beliefs about the value of the task (values) (Eccles & Wigfield, 2002). In EVT, the choice of task, performance on the task, and persistence on the task are affected by one's expectancies and values: All other variables and constructs that may have an effect on achievement are mediated through the expectancies and values constructs (Eccles, 1983; Eccles & Wigfield, 2002; Wigfield & Cambria, 2010). While a detailed description of the EVT models used by the SOMAS project team is beyond the scope of this chapter, a brief description of the rationale for selecting EVT is provided in the description of validity evidence section that follows.

The SOMAS project team began by developing EVT models that explain student performance based on the constructs and relationships articulated in the EVT literature (Eccles, 1983, 2014; Eccles & Wigfield, 2002). This led to the creation of the EVT model diagram for the S-SOMAS (Figure 6.2), which is similar to other published EVT models (e.g., Eccles, 2014). Each of the components of the model in Figure 6.2 (e.g., Utility Value, Goal Orientation) is a construct, a theoretical idea that we wish to measure (Wilson, 2005). The S-SOMAS model diagram, along with other internal documents describing the constructs, are heavily referenced by the SOMAS project team throughout the development process: it informs the item-writing process, the assembly of forms, analysis of data, and interpretation of results.

The model in Figure 6.2 was developed for use throughout the development process by illustrating all constructs hypothesized to influence performance in EVT, the relationships among these constructs, and which constructs are intended to be measured by the instrument. Further details about the EVT models are available in other manuscripts (e.g., Whitaker, Unfried, & Batakci, 2018).

Overview of Validity Evidence Plan

Validity, in the context of instrument development, refers to the overall support for proposed interpretations and uses of scores from an instrument (AERA et al., 2014; Messick, 1995). Validity is not an intrinsic property of an instrument; instead, it should always be discussed in the context of specific interpretations (AERA et al., 2014; Messick, 1995). It is only with appropriate evidence that specific interpretations and uses of instruments are supported, and collecting validity evidence is a key aspect of the instrument development process.

The SOMAS team has adopted the validity framework articulated in the 2014 *Standards for Educational and Psychological Testing* document (AERA et al., 2014) which conceptualizes as evidence supporting validity of interpretations as being derived from several sources. This framework describes five broad sources of evidence for validity claims:

- Evidence based on test content
- Evidence based on response processes
- Evidence based on internal structure
- Evidence based on relations to other variables
- Evidence for validity and consequences of testing

(AERA et al., 2014)

Evidence from each of these five sources supports intended interpretations and uses of an instrument (AERA et al., 2014).

Specific validity evidence to be collected aligned with each of the five sources supporting the use of the S-SOMAS instrument are presented below and summarized in Table 6.1. This table also lists the primary phase of the development process in which evidence will be collected. Note that the mapping of evidence statements to development phases is approximate and does not preclude evidence collection during other phases; for example, the claim "Chosen Likert-type response scale is appropriate" is listed as being a focus in the Initial Planning and Development phase, though evidence will also be collected in the Analysis of Collected Data and Item Writing and Revisions phases. Together, Table 6.1 and Figure 6.1 attempt to capture the complexity of the iterative instrument development process that includes the collection of many types of validity evidence both concurrently and sequentially. By focusing on validity evidence throughout the development process and modifying the instruments and supporting models as warranted by the development process, the resulting instruments will be well-supported for their intended uses by researchers with introductory statistics students.

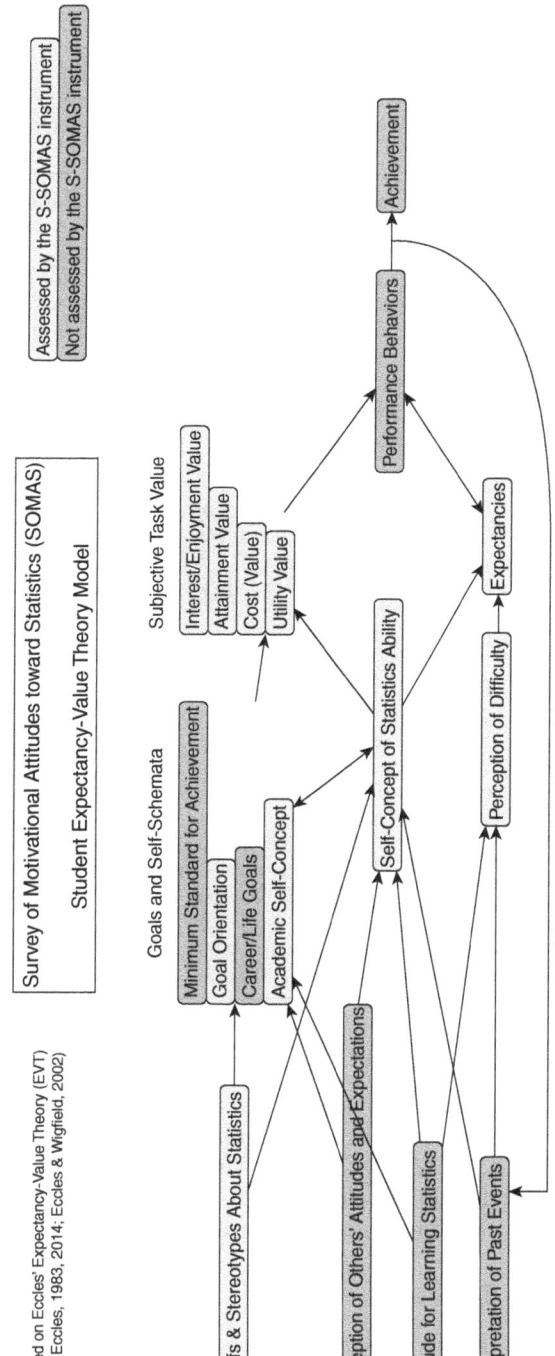

Figure 6.2 The student model based on expectancy-value theory.

Table 6.1 Claims supporting validity of interpretations for the S-SOMAS instrument for which evidence will be gathered as well as the primary development phase in which each statement will be the focus.

Source of Validity	Validity Evidence Claims	Primary Development Phase
Test content	The EVT model is appropriate for use with undergraduate students.	Initial planning and development
	Items are aligned with EVT constructs.	Item writing and revisions
	Created scales cover salient aspects of the constructs.	Assembly of forms and revisions
	Operationalized model is consistent with EVT model.	Assembly of forms and revisions
Response processes	Constructs have different levels on a continuum.	Analysis of collected data
	No differential item functioning will be present.	Analysis of collected data
	Chosen Likert-type response scale is appropriate.	Initial planning and development
	Online surveys are appropriate for data collection.	Initial planning and development
Internal structure	Items load onto hypothesized factors.	Item writing and revisions
	Each scale is internally consistent.	Analysis of collected data
	Measurement invariance for each construct.	Analysis of collected data
	Constructs are unidimensional.	Analysis of collected data
	Items are locally independent.	Analysis of collected data
Relations to other variables	Moderate positive correlations expected for the following factors.	Analysis of collected data
Uses and consequences	Appropriate for specific uses, inappropriate for others.	Intentional design and data collection throughout

Evidence Supporting Validity

Each of piece of validity evidence supporting the S-SOMAS development outlined in Table 6.1 is described and categorized using the five sources of validity evidence (AERA et al., 2014). These pieces of validity evidence serve as a roadmap for the development of the S-SOMAS instrument and may function as a model roadmap for future instrument development.

Validity Evidence Based on Test Content

The collection of validity evidence based on test content is central to the focus of developing an instrument to measure motivational attitudes toward statistics in a manner consistent with the established framework we have adopted. The evidence described below will be collected throughout the S-SOMAS development process, including the initial planning and development, item writing and revisions, and the assembly and revision of forms (see Table 6.1 and Figure 6.1). Clearly articulating these claims is integral to a plan that will result in an authentic assessment of attitudes toward statistics.

Claim: The EVT Model Is Appropriate for Use With Undergraduate Students

EVT was chosen as the theoretical framework to guide the development process, and more details about the selection and components of the proposed EVT models are planned for other manuscripts. The EVT model was initially considered as the theoretical framework for the S-SOMAS in part because it is the stated theoretical framework for the SATS-36 instrument, but there were additional reasons why the EVT model was considered and finally adopted for use in this project. The statistics education literature currently uses the language of attitudes, but it seems that researchers are interested in motivation to explain performance and behaviors. This led to the consideration of several motivational models.

Bandura's (1977, 1986) self-efficacy model serves as a theoretical foundation for EVT and other models such as self-regulated learning. Ultimately, these models do not conflict with each other. The use of Bandura's self-efficacy model directly was considered, but ultimately EVT was preferred because it includes additional aspects of motivation beyond self-efficacy. Self-regulated learning (Zimmerman & Labuhn, 2012) was not chosen because it is a cyclical model that would be difficult to fully investigate using a survey in one or two administrations. Ultimately, EVT (Eccles, 1983, 2014; Eccles & Wigfield, 2002) was chosen because of the many factors that it conceptualizes as influencing motivation while being consistent with other respected, widely used theories such as Bandura's self-efficacy model.

Another consideration that led to the adoption of EVT was that a single underlying theoretical framework explaining both students' motivations for learning statistics (using the S-SOMAS) and instructors' motivations for teaching statistics (using the I-SOMAS) was desired. Initial EVT models were developed for use with students who were children or adolescents (Eccles, 1983; Eccles & Wigfield, 1995). The implicit claim that the EVT model was appropriate for use with the populations of interest—undergraduate students and instructors—is made explicit.

Evidence supporting the appropriateness of EVT beyond the populations was initially proposed comes primarily from the literature. First, there is widespread consensus within the statistics education literature that EVT is appropriate for use with undergraduate statistics students as evidenced by the widespread use of the SATS-36 instrument and development of attitudinal models consistent with EVT (Ramirez et al., 2012; Schau, 2003b; Schau et al., 2012). Two empirical studies have found at least partial support for the use of EVT in statistics education, and discrepancies might be attributable to the instrument used in the studies (Hood, Creed, & Neumann, 2012; Sorge & Schau, 2002). Second, this EVT model has been widely applied with adults in various contexts, for example, with Korean female college students (Bong, 2001) and unemployed Belgians (Vansteenkiste, Lens, Witte, & Feather, 2005). Lastly, though a detailed argument based on empirical research about the appropriateness of the model with adults was not found, when Eccles and her colleagues write about EVT without a context, the term *individual* is used rather than *adolescent* or *child* (e.g., Eccles, 2014; Eccles & Wigfield, 2002). Even though Eccles and her colleagues have tended to apply the EVT model with children and adolescents, the widespread use of EVT with adults in the literature supports our use of EVT with adult populations, including undergraduate students and instructors.

Additionally, throughout the instrument development process, the proposed EVT models will be re-evaluated based on data collection and revised as needed. The goal of these revisions is to enact a model that is consistent with EVT while being responsive to the context of learning and teaching statistics. However, these models will ultimately be empirically tested once instruments have been constructed that will further confirm—or disconfirm—the appropriateness of the EVT model with undergraduate students and instructors.

Claim: Degree to Which Items Are Aligned With EVT Constructs

The items that comprise the SOMAS instruments are claimed to align with the EVT constructs in the theoretical framework. This claim is distinct from the loading of items onto statistical factors as discussed in the Internal Structure section: This claim is that the items themselves use language and ideas that are consistent with the theoretical constructs as defined in EVT rather than other frameworks.

The evidence supporting this claim is based on the item-writing and revision process used throughout the project. EVT was chosen as a theoretical framework and specific models for students and instructors were proposed prior to writing items or considering existing items for the SOMAS instruments. In this way, the SOMAS development team became familiar first with EVT and then items were written rather than doing a post-hoc alignment. Item writers consisted of the SOMAS development team and several members of ROSA. To guide the item-writing process, a document with directions, working definitions, and examples was created after the EVT models were proposed. This document was circulated to the item writers so that there were common touchstones all could reference. The working definitions were brief (a few sentences at most) and grounded in the literature, and the examples were a few statements for each construct that the development team believed would help clarify the associated definition (see Figure 6.3). At this point, item writers worked individually.

After items had been written, the SOMAS development team met to evaluate the alignment with constructs to determine an initial item pool for the S-SOMAS. Each item was evaluated for its alignment with the construct it was written for and other characteristics (e.g., readability, not double-barreled, etc.). Many items were excluded or revised substantially, and then a round of collaborative item writing took place. The development team met again to assess item alignment and develop the initial item pool from the proposed items. During this process, the SOMAS team struggled with aligning items in the pool for several constructs and, in doing so, refined its understanding of the EVT constructs and model.

Additionally, subject-matter experts (SMEs) have been—and will continue to be—involved in the review of items. After the initial pool of items was developed, a list was developed consisting of SMEs with expertise in statistics education, STEM education, student attitudes, the SATS instrument, and educational psychology. SMEs were identified by SOMAS team members using personal contacts and by searching the literature and conference programs for authors whose work was related one of the desired categories. These SMEs were presented with all items in the pool, organized by construct, and asked to rate how essential the item was for measuring the construct (Essential; Useful, but not Essential; Not Necessary). SMEs were also asked for their feedback about the overall construct, and many SMEs used this free-response box to discuss perceived item (mis) alignment. Both the qualitative and quantitative feedback was reviewed by the SOMAS project team and used to identify items to remove, recategorize, or revise. This differs from a Delphi study wherein participants are presented with the results from the group before providing their feedback again (Helmer-Hirschberg, 1967). The SOMAS project team rather than

---

Self-Concept of Statistics Ability

Statements concerning a student's concept of who they are in the domain of statistics. This is a component of academic self-concept but that specifically concerns students' perceptions of who they are in the domain of statistics. (Shavelson & Bolus, 1982, p. 3).

Examples: I can complete statistics problems because I am good at statistics. I have trouble understanding statistics because of how I think. I am good at statistics.

Figure 6.3 An example of the guidelines given to item writers.

the SMEs received the initial feedback. A similar process will be used in other phases of the project.

Claim: Created Scales Cover Salient Aspects of the Constructs

When developing instruments that are aligned with EVT, an implicit claim is that the scales that comprise the instrument measure the salient aspects of the constructs to which they are aligned. While all constructs are conceptualized as unidimensional continua, it is still not guaranteed that any set of items that have been separately identified as aligning to the construct and that load onto the statistical factor together capture the richness of the construct (because important aspects may not be assessed by any of the items). For example, the Utility Value construct in the student EVT model is the value one places on statistics because learning statistics meets some future goal (Eccles & Wigfield, 2002; Flake, Barron, Hulleman, McCoach, & Welsh, 2015). It is conceivable that a scale could be developed that measures the Utility Value construct in a limited way that ignores import some aspects, for example, valuing statistics because learning statistics specifically meets the future goal of finding a job (when really there are many future goals one might have that cause one to value statistics).

Evidence supporting this claim will be collected from a review by SMEs with expertise in EVT. These experts will be asked whether they believe the scales measure the salient aspects of the constructs, if any important aspects have been missed, or if there are irrelevant aspects that have been included. SMEs will be identified using a process similar to the one described above. Additionally, inter-item correlations will be calculated within scales to identify potential redundancies or scales that might be too homogenous and thus candidates for further review. However, this analysis is not sufficient for identifying when salient aspects have been missed.

Focus groups with various stakeholders are also planned to further support a robust view of each construct and have already been conducted with undergraduate students. Focus groups will also be held with instructors about the S-SOMAS instrument because they are an intended user of the instrument. These focus groups will be conducted early in the development process to inform item writing and construct alignment rather than centered on near-final instruments.

Claim: Operationalized Model Is Consistent With EVT Model

EVT is an adaptable framework: The full model has many constructs and relationships (Eccles, 2014; Eccles & Wigfield, 2002), while specific uses of EVT may make adaptations (e.g., Eccles, 2014), and a given instrument may not measure every construct. While EVT is adaptable, modifications and simplifications can threaten the extent to which an instrument is consistent with EVT. By using EVT as the theoretical framework for this family of instruments and developing models aligned with it, an explicit claim is made that the final, operationalized models are consistent with EVT.

As with assessing the scales for alignment with and coverage of EVT constructs described above, SMEs with expertise in EVT will be asked to provide feedback about the proposed EVT models and any simplifications, adaptations, or decisions about measurement that have been made to ensure that important aspects have not been sacrificed for some other end. This feedback will be solicited prior to finalizing instruments to allow time to respond to feedback and modify the models as-needed.

Claims Requiring Evidence Based on Response Processes

One definition of response processes is "the mechanisms that underlie what people do, think, or feel when interacting with, and responding to, the item or task" (Hubley & Zumbo, 2017, p. 2). There are many ways of collecting validity evidence for response processes (e.g., Padilla & Benítez, 2014), which include but are not limited to focus groups, interviews, and analyzing other specialized data from participants such as response times or eye movements. Hubley and Zumbo (2017) note that response process validity evidence may either be descriptive of how participants respond or explore why people respond in the ways that they do more deeply, and that while reporting of all types of response process validity evidence is limited, most work has been descriptive. The claims presented below for the SOMAS project are descriptive in the sense offered by Hubley and Zumbo. Validity evidence supporting these claims about the ways in which respondents interact with the instrument will be collected during planning and development phases as well as when the assessments are in pilot and operational phases (see Table 6.1 and Figure 6.1). Because validation is never complete, these descriptive claims are for initial response process validity evidence, but future work may expand to further explore why participants answer in the ways that they do. Like many instruments used in education, the S-SOMAS is an online survey. However, this does not abrogate the need to collect validity evidence based on response processes and plan accordingly.

Claim: Constructs Have Different Levels on a Continuum

Consistent with Coombs' formulation that endorsing a response on a Likert-type item is based on the relative location of an individual and an item on the underlying continuum for that construct (1964; Roberts, Laughlin, & Wedell, 1999), each construct in the EVT model is conceptualized as being a continuum, and each respondent will have some level of the construct on that continuum. The probability of endorsing a particular response in a Likert-type item is therefore based on an individual's location on an underlying continuum, and it is this hypothesized continuum that is the focus of this claim. For the S-SOMAS instrument, this continuum will be manifest in the ordering of respondents based on their responses to the items in the scale rather than an ordering of items within each scale along a continuum (Wilson, 2005). For example, for the utility value construct items will not be written that have an explicit order and mapping from low utility value to high utility value along a continuum. Instead, we expect to see students respond in such a way that the scale scores for students fall along a continuum from having low to high utility value for statistics. The primary way that validity evidence for this claim will be collected is the construction of a Wright map (Wilson, 2005, 2011) during analysis of data collected from pilot and operational administrations of the S-SOMAS. A Wright map, sometimes called an item-person map, is a figure illustrating the item difficulty continuum and person ability continuum on the same axis and scale (Wilson, 2011). It is hypothesized that the scales measuring EVT constructs will produce Wright maps that show respondents at various levels of the construct, thus illustrating some continuum for the construct that influences an individual's response process.

Claim: No Differential Item Functioning Will Be Present

The SOMAS instruments will also be investigated for the possibility of differential item functioning (DIF), which occurs when items perform differently for respondents of the same level on the continuum of interest who belong to different groups. DIF is undesirable and reflects items performing differently across groups rather than underlying group differences (Wilson, 2005). Initially, DIF will only be investigated between groups of respondents who self-identify as female or male on the student instrument due to sample size requirements. However, as more data are collected DIF will be assessed for other identifiable groups (e.g., based on self-identified race). While the initial administrations of the instruments will be in the United States, it is hoped that the instruments will be used broadly: As uses and data support such analyses, DIF will be investigated across global cultures.

Claim: Chosen Likert-Type Response Scale Is Appropriate

All items on the SOMAS instruments will use a 7-point Likert-type response scale (Strongly Disagree to Strongly Agree with a Neutral category). There are two claims related to the selection of the 7-point scale that must be justified: (1) the 7-point scale provides enough granularity to account for respondents across the hypothesized continuum for each construct and (2) the inclusion of a midpoint is appropriate. Evidence for both claims is drawn primarily from the literature rather than empirically derived using data collected during this project.

The use of a midpoint in the Likert-type items is supported because a neutral response might be theoretically more appropriate for a respondent on any given item. Additionally, the choice to use or not use a neutral response category has not been shown to have a substantial effect on data or conclusions (Dillman, Smyth, & Christian, 2014). For Likert-type items, 4 to 7 points is a widely used guideline (Dillman et al., 2014). In the statistics education literature, 7-point scales are widely used and selecting the same scale may aid in simplifying interpretations and comparisons of instruments.

Claim: Online Surveys Are Appropriate for Data Collection

An implicit claim made of the SOMAS instruments is that an online administration format is appropriate. The use of online survey tools is ubiquitous in contemporary social science research (Dillman et al., 2014; Hewson & Stewart, 2016), and the populations of interest—students and instructors of statistics—seem well-suited to web-based surveys. Concerns about the use of internet surveys with certain populations generally stems from potential added costs, lack of access, or low computer literacy (Dillman et al., 2014). Students and teachers of statistics have increasingly used computers for decades (suggesting at least basic computer skills for most students and their teachers), and the use of technology in statistics is a widely accepted best practice (GAISE College Report ASA Revision Committee, 2016).

Additional concerns about low response rates to internet surveys are not generally applicable to the use of the SOMAS instruments because the anticipated mode of administration is for students to be solicited for participation by their instructor, an individual known to them rather than a stranger. There is a potential for nonresponse rates for instructors to be high because the researchers may be strangers to them, but the statistics education community is friendly and targeted efforts will be made to recruit statistics instructors.

Potential trust issues related to the collection and storage of data will also be mitigated using a professional survey administration platform. Initially, Qualtrics (a professional platform with many options for securely storing data and ensuring respondents can use a multitude of devices to respond) will be used for the S-SOMAS with a goal of later moving to a customized, professional website modeled on existing successful websites used for data collection in statistics education. We believe that this approach for recruiting participants and collecting data ameliorates many of the challenges associated with using internet surveys and that the benefits greatly outweigh the challenges.

Claims Requiring Evidence Based on Internal Structure

The validity evidence based on internal structure is closely related to properties of the instrument, scales, and items, and primarily a focus during the analysis of data collected from pilot and operational administrations of the instrument, though some validity evidence stems from the item writing and revision process (see Table 6.1 and Figure 6.1). While validity evidence based on internal structure is among the most common reported (Hubley & Zumbo, 2017), planning for the analyses required and understanding how they will affect revisions to instruments and models is key.

Claim: Items Load onto Hypothesized Factors

A particularly important claim that is made about the SOMAS assessments is that the assignment of items to scales is appropriate. That is, items presented as part of a scale designed to measure an underlying construct should measure that construct and not another. While this claim may seem to be easily met on its surface, early item-writing efforts based on the initial S-SOMAS EVT model revealed unforeseen similarities among items written for different constructs. For example, when reviewing items written to measure the Goal Orientation construct and the Attainment Value construct, the research team experienced difficulties determining for an individual item which construct it actually measured. An example of the similarity among items from different constructs can be seen in these provisional items, each from a different construct:

- If I am unable to interpret statistical results, I feel insecure. (Attainment Value)
- It is challenging to solve a problem that requires using statistics. (Perceived Difficulty)
- I can complete tasks that require basic statistical skills. (Expectancies)
- I lack the skills to do well in statistics. (Self-Concept of Statistics)

Each of the four items above was written for and aligns with the construct indicated in parentheses but viewed together there are considerable similarities. In response to this difficulty, the complete pool of items as of Fall 2017 was administered to students enrolled in introductory statistics courses in two forms composed of scales designed to measure constructs that were perceived as similar by the research team. All items were included on at least one form, neither form contained items from all constructs, and the form that an individual received was randomized. The purpose of this early pilot administration (Pilot 1 in Figure 6.1) is to provide feedback to the SOMAS team about the nature of the items and constructs. The data collected will be analyzed using exploratory factor analysis with a focus on item loadings: items with high loadings in only one construct are desired for this project. The analysis of data collected in this pilot administration will inform the assignment of items to scales, the item-writing process, and potential revisions to the EVT models. Throughout development factor loadings for items will be examined to determine if the item is appropriately categorized.

Claim: Each Subscale Is Internally Consistent

Related to the above claim is the claim that the final scales on the SOMAS instruments will have reasonable internal consistency, that is, "the degree to which the items on a test jointly measure the same construct" (Henson, 2001, p. 177). However, while internal consistency is an important aspect of reliability, particularly high values for measures of internal consistency for each scale are not of paramount importance for this project. This is because measures of internal consistency tend to be higher for longer scales, and the number of constructs to be measured would result in prohibitively long S-SOMAS and I-SOMAS instruments that would be unlikely to see widespread use. Depending on the specific development of scales, longer scales measuring a construct may be developed and then pared down to produce shorter scales that still have acceptable internal consistency.

The omega coefficient (Raykov & Marcoulides, 2011) will be used as the primary measure of internal consistency of scales on SOMAS instruments to be consistent with the confirmatory factor analysis approach to be used when finalizing instruments. Other measures of internal consistency—such as coefficient alpha (Cronbach, 1951)—may be calculated and used throughout the project for communicating with other audiences but only with an understanding of their uses and limitations (e.g., Henson, 2001; Sijtsma, 2009). Because of coefficient alpha's ubiquity, a few limitations will be briefly discussed.

Sijtsma (2009) provides a thorough description of coefficient alpha and details several fundamental problems with its typical use. First, Sijtsma notes that alpha is grounded in the paradigm of classical test theory, and so its ad-hoc use with other paradigms such as item response theory is not advisable. Second, while alpha is correlated with other statistical measures, it is neither a measure of internal consistency nor a measure of unidimensionality and conveys little information on its own (Sijtsma, 2009). Third, alpha's use as a lower-bound estimate for an instrument's reliability (its correlation between the scores on an instrument an on a parallel version) based on a single administration is often inappropriate because better lower bounds have been proposed (Sijtsma, 2009). Lastly, a focus on reporting alpha may serve to oversimplify and conflate the distinct concepts of reliability and internal consistency.

Claim: Measurement Invariance for Each Construct

Another claim that evidence will be collected is that the measurement is invariant for each construct, that is, the distances between respondents and the distances between responses on a Wright map are interpretable regardless of location on the continuum (Wilson, 2003, 2005). Evidence related to this claim will be gathered during data analysis: measurement invariance equates to each item having the same slope in the measurement model that is adopted. Because of the 7-point Likert-type scale that is being used with the S-SOMAS instrument, a polytomous model such as the generalized partial credit model (Muraki, 1992) or graded response model (Samejima, 1969) will be employed during the analysis of data. Models for analyzing polytomous responses differ in the parameters that are estimated. While a comparison of each model is beyond the scope of this chapter, one way in which these models differ is whether the model for all items has the same slope (discrimination) or if each item has a slope parameter calculated. Together with the researchers' judgment of the appropriateness of the models, statistical techniques will be used to compare models to determine if equality of slope parameters for each item is reasonable.

If a model that allows each item to have a different slope parameter fits best, then two options may be considered. First, the items that measure the construct under consideration will be revisited, and items may be added or deleted to obtain a measure of the construct that is invariant (Wilson, 2005). In this case, the items with slope parameters most dissimilar from the others would be examined first to determine if they are candidates for removal; the difference in item slope parameters may be a determining factor in the deletion of an item in addition to other reasons. If this is not successful, then the interpretations of the construct under consideration may be revised to account for this invariances through a more complex interpretation (Wilson, 2005). The first option is preferable for this project because the final family of instruments are intended to be made widely available to researchers, instructors, and other interested parties, and parsimonious construct interpretations may be more desirable than complex ones. However, proposed interpretations of constructs may be made more complex if needed (and supplemented with appropriate training materials).

Claim: Constructs Are Unidimensional

We plan to use item response theory (IRT) to analyze responses, and IRT assumes that underlying constructs measured by each scale are unidimensional. This is consistent with the EVT framework guiding the development. There are two related claims about unidimensionality for this project: (1) that the EVT constructs to be measured by the SOMAS instruments are unidimensional and (2) that a unidimensional structure is appropriate for each scale created to measure an EVT construct.

Each EVT construct proposed by Eccles and colleagues (e.g., Eccles & Wigfield, 2002) is not explicitly unidimensional, though many of the constructs are expected to be unidimensional when operationalized. For example, the EVT construct Utility Value, in the context of learning statistics, is expected to be a unidimensional construct: The SATS-36 Value construct is most closely aligned with Utility Value (Whitaker & Gorney, 2017) and has been shown to be a distinct unidimensional factor based on confirmatory factor analysis (Vanhoof et al., 2011).

It is likely, though, that some proposed EVT constructs will not be facially unidimensional. During an early item-writing activity, the Goal Orientation construct was identified as being potentially multidimensional. Initially, a unidimensional continuum of intrinsic orientation to extrinsic orientation was assumed. However, difficulties in writing and reviewing items led to the decision to view Goal Orientation as two distinct but related unidimensional constructs: Intrinsic Goal Orientation and Extrinsic Goal Orientation, each on a low to high continuum. When data suggest that a construct may not be unidimensional and a theoretical explanation can be supported, broader constructs may be reconceptualized as narrower unidimensional constructs. The Intrinsic and Extrinsic Goal Orientation constructs reflect different orientations that have been proposed in the literature such as mastery and performance orientation, respectively (Wigfield & Cambria, 2010). The proposed EVT models will then be refined.

Moreover, it is possible that some proposed EVT constructs might not be practically distinguishable from each other and therefore best viewed not as distinct unidimensional constructs but rather as a single unidimensional construct. Eccles and Wigfield (2002) define expectancies for success by synthesizing the extant work of Eccles and her colleagues: "individuals' beliefs about how well they will do on upcoming tasks, either in the immediate or longer term future" (p. 119). While Eccles and Wigfield include two different theoretical types of beliefs in their definition—beliefs about the immediate future and beliefs about the longer-term future—they note that these "are highly related and empirically indistinguishable" (p. 119). When data suggest that theoretically distinct constructs may not be empirically distinguishable and a theoretical explanation can be supported, narrower constructs may be collapsed to form broader unidimensional constructs. The proposed EVT models will then be refined.

Evidence about unidimensionality will be gathered throughout the project. First, as items are written for specific constructs and preliminary data collected, exploratory factor analysis will be used to guide and refine the assignment of items to constructs. Additionally, items will be rewritten, omitted, or replaced as needed to support the creation of scales that perform in ways consistent with measuring unidimensional constructs. The creation of scales that measure unidimensional constructs in the EVT model will be an iterative process where the theoretical framework guides the creation of scales and data collected from the use of the scales informs revisions to the theoretical framework. Ultimately, prior to supporting the widespread use of the SOMAS instruments by other researchers, model fit in a confirmatory factor analysis setting will be assessed to support the unidimensionality of the constructs measured by the revised instruments aligned with revised EVT frameworks.

Claim: Items Are Locally Independent

Another assumption needed for IRT is local independence, that is, responses to items are independent given a respondent's individual trait characteristics. To assess local independence, pairwise inter-item correlations will be computed and examined. For at least one pilot administration, items will be administered either grouped by constructs and in a randomized order before a final choice is made.

Claims Requiring Evidence Based on Relations to Other Variables

As part of the initial validity study, other instruments will be administered simultaneously with the S-SOMAS instrument. There is a strong presence in the planning phases to determine which instruments to administer and to whom. These instruments will not all be administered to all participants. Instead, subsets of respondents will be selected to take the S-SOMAS instrument and another instrument to determine if hypothesized relationships are observed. Relationships between the S-SOMAS instrument and two other instruments will be the primary focus: the SATS-36 (Schau, 2003b) and the Levels of Conceptual Understanding in Statistics Intermediate/Advanced online form (LOCUS; Jacobbe, Case, Whitaker, & Foti, 2014; Whitaker, Foti, & Jacobbe, 2015). These hypotheses, which are grounded in the statistics education literature, motivate the data collection plan and gathering of validity evidence in a deliberate way rather than being incidental components of later research.

Because the SATS-36 and S-SOMAS instrument both aim to measure students' attitudes toward statistics in a manner consistent with EVT, positive correlations between comparable constructs are expected. In particular, we would expect to see moderate positive correlations for the following factors:

- SATS-36 Affect and S-SOMAS Subjective Task Value subconstructs
- SATS-36 Cognitive Competence and S-SOMAS Self-Concept of Statistics Ability
- SATS-36 Value and S-SOMAS Utility Value
- SATS-36 Difficulty and S-SOMAS Perception of Difficulty
- SATS-36 Interest and S-SOMAS Interest/Enjoyment

These hypothesized relationships are based on the existing statistics education literature using the SATS instruments (e.g., Schau, 2003a; Schau & Emmioğlu, 2012; Vanhoof et al., 2011). Additionally, we do not expect to see a moderate correlation between SATS-36 Effort and S-SOMAS Cost because the SATS-36 Effort construct has previously been noted to produce responses that might be too high to reflect reality (Schau & Emmioğlu, 2012). Other weak to moderate correlations among the constructs are expected to be observed, but these predictions are articulated because the scales on each instrument are attempts to measure similar underlying constructs.

Evidence for Validity and Consequences of Testing

Validity evidence will also be collected to support specific uses of the S-SOMAS. The developers intend that these instruments will be appropriate for longitudinal and pre/post-research designs as well as snapshots of motivational attitudes with students and instructors at higher education institutions within the United States. To support this broad claim, several claims have been articulated above, and these specific research purposes precipitated the data collection plan for the S-SOMAS. Initial validation work will use the S-SOMAS instruments in a pre/post design, as that has been popular in statistics education and was the intended use of the SATS-36, but S-SOMAS is being designed so that researchers are not restricted to only using it in a pre/post setting but instead are appropriate for use longitudinally to support more sophisticated research designs such as latent growth models and time series (e.g., Sloane & Wilkins, 2017). Additionally, data collection will include a diverse set of higher education institutions in the United States, such as large public research universities; small liberal arts schools, primarily undergraduate institutions; and community colleges.

While the SOMAS development team does not anticipate uses of the instruments beyond research in statistics education, to clarify the claims about intended uses an explicit anti-endorsement of use is offered: the SOMAS instruments should not be used in high-stakes settings such as hiring or firing instructors or placing students into courses. There are likely other uses for SOMAS instruments that might be appropriate that have not been articulated by the development team, but these other uses would require explicit validity evidence to be collected to support them and cannot rely only on the validity evidence collected by the development team.

Use of the Validity Plan in Guiding Development

This chapter aimed to illustrate the development process for an instrument to be used for research purposes with a focus on validation work with explicit attention to several aspects that may be hidden from or unknown to those new to instrument development: the development process is not linear—revisions occur in many areas, including in the documents guiding the development as more is learned—and intended uses are supported by a variety of claims for which evidence must be collected. Each preceding claim, along with its potential evidence, was written intentionally to support the SOMAS project's goals and the intended use of the S-SOMAS with undergraduate students enrolled in statistics courses. While some claims may be made in many validation studies (e.g., "Items are locally independent" is a requirement of IRT), each claim was made to support specific uses of the S-SOMAS for specific purposes. The ultimate goal of the SOMAS project is not to develop an instrument: We want to deepen the field's knowledge of what affects student outcomes in statistics and to what degree. To that end, a new family of instruments is needed as there is a paucity of validity evidence for existing instruments supporting our desired uses. Our intended uses of the S-SOMAS instrument motivate the development of the instrument and the validity evidence to be collected.

As illustrated in Figure 6.1, the development of the S-SOMAS instrument is underway but with much work remaining. As of November 2018, the EVT framework has been developed and revised multiple times, a pool of items has been written and revised, SMEs have been consulted, focus groups have been conducted, and data have been collected in a pilot administration to inform revisions. While this is still relatively early in the overall conception of the larger SOMAS project, we have already needed many of the validity claims listed above for guiding and focusing our discussions and research activities. For example, prior to starting any item writing, some evidence for the following validity claims was needed:

- The EVT model is appropriate for use with students,
- The chosen Likert-type response scale is appropriate, and
- Online surveys are appropriate for data collection.

Additionally, the validity statements concerning appropriate use of the final instruments were also critical to have articulated before item writing began. The validity statements serve to focus the work done by the SOMAS project team and to help coordinate the many research tasks that will be undertaken in the coming years.

Conclusion and Discussion

Developing high-quality instruments for which there is documentable evidence supporting their intended uses requires substantial planning and commitment. The SOMAS project is an ongoing instrument development project, and specific details and goals described in this chapter evolve or change as the project continues—but will be guided by the same focus on explicit evidence supporting intended uses that was demonstrated here. A timeline of the development process is illustrated in Figure 6.1; this timeline shows the sequencing of key development activities and highlights the iterative nature of the instrument development process: The revisions to the outcomes of previous phases of development are explicitly accounted for based on the collection and analysis of additional data. The process of developing a high-quality instrument that is suitable for its intended purpose takes substantial time: Developing an instrument is not just writing a survey or test, and it cannot be completed overnight or even in a few weeks. While many factors affect the amount of time needed to develop an instrument, planning on a timeframe that is measured in *months* rather than *days* is advisable.

The development of the S-SOMAS instrument exemplifies what this planning can look like with a focus on the collection of validity evidence consistent with the five sources of validity evidence in the *Standards for Educational and Psychological Testing* (AERA et al., 2014). The validity evidence described in this chapter was summarized in Table 6.1. Together, Table 6.1 and Figure 6.1 attempt to capture the complexity of a multiyear instrument development project. By making explicit validity evidence that will support the intended uses of the S-SOMAS instrument early in the development process, these evidence statements serve to guide development toward the intended uses rather than resulting in a piecemeal approach to validity undertaken after development has concluded.

References

American Educational Research Association, American Psychological Association, & National Council on Measurement in Education. (2014). *Standards for educational and psychological testing.* Washington, DC: American Educational Research Association.

Bandura, A. (1977). *Social learning theory.* Englewood Cliffs, NJ: Prentice Hall.

Bandura, A. (1986). *Social foundations of thought and action: A social cognitive theory.* Englewood Cliffs, NJ: Prentice-Hall.

Bong, M. (2001). Role of self-efficacy and task-value in predicting college students' course performance and future enrollment intentions. *Contemporary Educational Psychology, 26*(4), 553–570.

Coombs, C. H. (1964). *A theory of data.* Oxford, England: Wiley.

Cronbach, L. J. (1951). Coefficient alpha and the internal structure of tests. *Psychometrika, 16*(3), 297–334.

Dauphinee, T. L., Schau, C., & Stevens, J. J. (1997). Survey of attitudes toward statistics: Factor structure and factorial invariance for women and men. *Structural Equation Modeling: A Multidisciplinary Journal, 4*(2), 129–141.

Dillman, D. A., Smyth, J. D., & Christian, L. M. (2014). *Internet, phone, mail, and mixed-mode surveys: The tailored design method* (4th ed.). Hoboken: Wiley.

Eccles, J. S. (1983). Expectancies, values, and academic behaviors. In J. T. Spence (Ed.), *Achievement and achievement motives: Psychological and sociological approaches* (pp. 75–145). San Francisco: W.H. Freeman.

Eccles, J. S. (2014). Expectancy-value theory. In R. Eklund & G. Tenenbaum (Eds.), *Encyclopedia of sport and exercise psychology*. Thousand Oaks, CA: Sage Publications, Inc.

Eccles, J. S., & Wigfield, A. (1995). In the mind of the achiever: The structure of adolescents' academic achievement related-beliefs and self-perceptions. *Personality and Social Psychology Bulletin, 21*(3), 215–225.

Eccles, J. S., & Wigfield, A. (2002). Motivational beliefs, values, and goals. *Annual Review of Psychology, 53*, 109–132.

Flake, J. K., Barron, K. E., Hulleman, C., McCoach, B. D., & Welsh, M. E. (2015). Measuring cost: The forgotten component of expectancy-value theory. *Contemporary Educational Psychology, 41*, 232–244.

GAISE College Report ASA Revision Committee. (2016). *Guidelines for assessment and instruction in statistics education college report 2016*. Retrieved from https://www.amstat.org/asa/files/pdfs/GAISE/GaiseCollege_Full.pdf

Gal, I., Ginsburg, L., & Schau, C. (1997). Monitoring attitudes and beliefs in statistics education. In I. Gal & J. B. Garfield (Eds.), *The assessment challenge in statistics education* (pp. 37–51). Amsterdam: IOS Press.

Helmer-Hirschberg, O. (1967). *Analysis of the future: The Delphi method*. Santa Monica, CA: RAND Corporation.

Henson, R. K. (2001). Understanding internal consistency reliability estimates: A conceptual primer on coefficient alpha. *Measurement and Evaluation in Counseling and Development, 34*(3), 177–189.

Hewson, C., & Stewart, D. W. (2016). Internet research methods. In N. Balakrishnan, T. Colton, B. Everitt, W. Piegorsch, F. Ruggeri, & J. L. Teugels (Eds.), *Wiley StatsRef: Statistics reference online* (pp. 1–6). Chichester, UK: John Wiley & Sons, Ltd.

Hood, M., Creed, P. A., & Neumann, D. L. (2012). Using the expectancy value model of motivation to understand the relationship between student attitudes and achievement in statistics. *Statistics Education Research Journal, 11*(2), 72–85.

Hubley, A. M., & Zumbo, B. D. (2017). Response processes in the context of validity: Setting the stage. In B. D. Zumbo & A. M. Hubley (Eds.), *Understanding and investigating response processes in validation research* (pp. 1–12). Cham: Springer.

Jacobbe, T., Case, C., Whitaker, D., & Foti, S. (2014). Establishing the validity of the LOCUS assessments through an evidenced-centered design approach. In K. Makar & R. Gould (Eds.), *Proceedings of the ninth international conference on teaching statistics (ICOTS9, July, 2014), Flagstaff, Arizona, USA*. Voorburg, The Netherlands: International Statistical Institute.

Messick, S. (1995). Validity of psychological assessment: Validation of inferences from persons' responses and performances as scientific inquiry into score meaning. *American Psychologist, 50*(9), 741.

Muraki, E. (1992). A generalized partial credit model: Application of an EM algorithm. *Applied Psychological Measurement, 16*(2), 159–176.

Nolan, M. M., Beran, T., & Hecker, K. G. (2012). Surveys assessing students' attitudes toward statistics: A systematic review of validity and reliability. *Statistics Education Research Journal, 11*(2), 103–123.

Padilla, J. L., & Benítez, I. (2014). Validity evidence based on response processes. *Psicothema, 26*(1), 136–144.

Ramirez, C., Schau, C., & Emmioğlu, E. (2012). The importance of attitudes in statistics education. *Statistics Education Research Journal*, *11*(2), 57–71.

Raykov, T., & Marcoulides, G. A. (2011). *Introduction to psychometric theory*. New York: Routledge.

Roberts, J. S., Laughlin, J. E., & Wedell, D. H. (1999). Validity issues in the Likert and Thurstone approaches to attitude measurement. *Educational and Psychological Measurement*, *59*(2), 211–233.

Samejima, F. (1969). *Estimation of latent ability using a response pattern of graded scores*. Richmond, VA: Psychometric Society.

Schau, C. (1992). *Survey of Attitudes Toward Statistics (SATS-28)*. Retrieved from http://evaluationandstatistics.com/

Schau, C. (2003a). *Students' attitudes: The "other" important outcome in statistics education*. Presented at the Joint Statistics Meetings (pp. 3673–3683), San Francisco, CA.

Schau, C. (2003b). *Survey of Attitudes Toward Statistics (SATS-36)*. Retrieved from http://evaluationandstatistics.com/

Schau, C., & Emmioğlu, E. (2012). Do introductory statistics courses in the United States improve students' attitudes? *Statistics Education Research Journal*, *11*(2), 86–94.

Schau, C., Millar, M., & Petocz, P. (2012). Research on attitudes towards statistics. *Statistics Education Research Journal*, *11*(2), 2–5.

Schau, C., Stevens, J., Dauphinee, T. L., & Vecchio, A. D. (1995). The development and validation of the survey of attitudes toward statistics. *Educational and Psychological Measurement*, *55*(5), 868–875. doi:10.1177/0013164495055005022

Sijtsma, K. (2009). On the use, the misuse, and the very limited usefulness of Cronbach's alpha. *Psychometrika*, *74*(1), 107.

Sloane, F. C., & Wilkins, J. L. M. (2017). Aligning statistical modeling with theories of learning in mathematics education research. In J. Cai (Ed.), *Compendium for research in mathematics education* (pp. 183–207). Reston, VA: National Council of Teachers of Mathematics.

Sorge, C., & Schau, C. (2002). *Impact of engineering students' attitudes on achievement in statistics*. Paper presented at the American Educational Research Association Annual Meeting, New Orleans, LA.

Vanhoof, S., Kuppens, S., Sotos, A. E. C., Verschaffel, L., & Onghena, P. (2011). Measuring statistics attitudes: Structure of the Survey of Attitudes Toward Statistics (SATS-36). *Statistics Education Research Journal*, *10*(1), 35–51.

Vansteenkiste, V., Lens, W., Witte, H., & Feather, N. T. (2005). Understanding unemployed people's job search behaviour, unemployment experience and well-being: A comparison of expectancy-value theory and self-determination theory. *British Journal of Social Psychology*, *44*(2), 269–287.

Whitaker, D., Foti, S., & Jacobbe, T. (2015). The levels of conceptual understanding in statistics project: Results of the pilot study. *Numeracy*, *8*(2), Article 4.

Whitaker, D., & Gorney, K. (2017). *Surveys of attitudes about statistics: An analysis of items*. Poster presented at the 39th Annual Conference of the North American Chapter of the International Group for the Psychology of Mathematics Education, Indianapolis, IN.

Whitaker, D., Unfried, A., & Batakci, L. (2018). A framework and survey for measuring students' motivational attitudes toward statistics. In M. A. Sorto, A. White, & L. Guyot (Eds.), *Proceedings of the tenth international conference on teaching statistics (ICOTS10, July, 2018)*. Voorburg, The Netherlands: International Statistical Institute.

Wigfield, A., & Cambria, J. (2010). Students' achievement values, goal orientations, and interest: Definitions, development, and relations to achievement outcomes. *Developmental Review*, *30*(1), 1–35.

Wilson, M. (2003). On choosing a model for measuring. *Methods of Psychological Research Online*, *8*(3), 1–22.

Wilson, M. (2005). *Constructing measures: An item response modeling approach.* Mahwah, NJ: Lawrence Erlbaum Associates.

Wilson, M. (2011). Some notes on the term: "Wright map." *Rasch Measurement: Transactions of the Rasch Measurement SIG American Educational Research Association*, *25*(3), 1331.

Zimmerman, B. J., & Labuhn, A. S. (2012). Self-regulation of learning: Process approaches to personal development. In K. R. Harris, S. Graham, T. Urdan, & J. Brophy (Eds.), *APA educational psychology handbook* (pp. 399–426). Washington, DC: American Psychological Association.

7 Measuring Self-Efficacy to Teach Statistics in Grades 6–12 Mathematics Teachers

*Leigh M. Harrell-Williams, Jennifer N. Lovett,
Lawrence M. Lesser, Hollylynne S. Lee,
Rebecca L. Pierce, Teri J. Murphy, and
M. Alejandra Sorto*

Introduction

Significant developments in the fields of mathematics and statistics education motivated the need to examine mathematics teachers' confidence to teach statistics. Over the last three decades, there have been calls to increase the emphasis on statistics in grades K-12 (National Council of Teachers of Mathematics [NCTM], 1989, 2000). Such increased emphasis has happened nationally with many states adopting the Common Core State Standards for Mathematics (CCSSM; National Governors Association Center for Best Practice & Council of Chief State School Officers, 2010) or modifying their standards to align with the CCSSM (Certica Solutions, 2018). Since many in-service and preservice teachers had minimal experience with statistics as K-12 students, and possibly in their own teacher preparation program, they may not feel confident to implement the new statistics standards (Lovett & Lee, 2017). In response to the increased emphasis on statistics, and teachers' potentially limited interaction with the content, mathematics and statistics educators have designed professional development opportunities to enhance teachers statistical knowledge and confidence to teach statistics (e.g., Bargagliotti et al., 2014; Gould, Bargagliotti, & Johnson, 2017; Jones, Lovett, & Google, 2017; Lee & Stangl, 2015; Peters, Watkins, & Bennett, 2014). To enhance teacher education programs, the *Statistical Education of Teachers (SET)* report (Franklin et al., 2015) provided grade level–specific recommendations for the content and number of courses to develop both statistical content and pedagogical knowledge. To support evaluation of such efforts, instruments were needed to capture the statistical knowledge and confidence of teachers. Thus, this chapter discusses a pair of instruments designed to measure middle school and high school teachers' confidence to teach statistics, the studies that have been conducted with the instruments, how the tenets of Rasch measurement theory and Messick's (1995) validation framework informed instrument development and revision activities, and ties Messick's framework to the recently updated *Standards for Educational and Psychological Testing* (American Educational Research Association, American Psychological Association, & National Council on Measurement in Education [AERA, APA, & NCME], 2014).

Background Literature

Teaching Efficacy

Mathematics teachers hold beliefs about the nature of mathematics, about themselves as a learner of mathematics, their students, and student learning (Calderhead, 1996; Pierce & Chick, 2011). These beliefs influence their actions in the classroom and student achievement (Love & Kruger, 2005; Philipp, 2007; Wilkins, 2008; Zee & Koomen, 2016). A major component of teachers' belief systems is their self-efficacy for specific tasks associated with teaching (McGee & Wang, 2014). Self-efficacy is defined as "people's beliefs in their capabilities to produce given attainments" (Bandura, 2006, p. 307). One's self-efficacy is dynamic and task specific (Bandura, 1977; Pajares, 1997). Thus, mathematics teachers have self-efficacy to: (1) learn the content themselves and (2) teach the topic to students. The latter is known as teaching efficacy—a teacher's "belief that they have the skills to bring about student learning" (Ashton, 1985, p. 142).

Two instruments designed to measure mathematics teaching efficacy are the Mathematics Teaching Efficacy Beliefs Instrument (MTEBI; Enochs, Smith, & Huinker, 2000) and the Self-Efficacy for Teaching Mathematics Instrument (SETMI; McGee & Wang, 2014). The MTEBI was the first mathematics-specific instrument to measure mathematics teaching efficacy for which validity evidence was compiled. However, the items are more general in nature and do not focus on specific mathematical topics (e.g., "I will continually find better ways to teach mathematics" and "I am typically able to answer students' mathematics questions"). The SETMI was developed to provide a more topic-specific instrument and focuses on the elementary grades' topics of integers, rational numbers, irrational numbers, probability, size, quantity, and capacity. This instrument moved the field of mathematics teaching efficacy forward but is appropriate only to measure the mathematics teaching efficacy of elementary teachers. The field of mathematics teacher education clearly needed additional topic-specific instruments for middle school and high school mathematics teachers, as well as for additional content areas that mathematics teachers are expected to teach.

Statistics Teaching Efficacy

Statistics and probability comprise one of the domains of middle school and high school mathematics identified by CCSSM. However, there is a difference between mathematical reasoning and statistical reasoning within a mathematics classroom (Rossman, Medina, & Chance, 2006). delMas (2004) describes mathematics as "the study of patterns; therefore, mathematical reasoning involves reasoning about patterns" (p. 82). Mathematical reasoning often removes the context to find patterns (Rossman et al., 2006); whereas statistical reasoning also looks for patterns but the meaning of the patterns depends on the context of the data (Cobb & Moore, 1997) and conclusions always involve uncertainty (Rossman et al., 2006). Statistical thinking involves engaging in a four-phase statistical investigative cycle (Franklin et al., 2007): formulate statistical questions, collect data, analyze data, and interpret results. Considering these differences in mathematics and statistics, K-12 mathematics teachers may hold different beliefs about their ability to teach statistics content as compared to mathematics content. These beliefs are known as their statistics teaching efficacy; "a teacher's belief in his/her ability to teach statistics to bring about student learning" (Lovett, 2016, p. 83). Prior to the creation of the instruments described in this chapter, only a few studies had been conducted on the statistics teaching efficacy of preservice or in-service teachers. When examining the statistics teaching efficacy of preservice teachers through interviews, Fitzmaurice, Leavy, and Hannigan (2014) found that preservice teachers were reluctant about having to teach statistics during their student teaching and did not feel confident to do so. However, those who were required to teach statistics during student teaching later reported increased statistics teaching efficacy. Watson (2001) designed a teacher-profiling instrument that was administered as an interview or questionnaire to a group of 43 practicing teachers across K-12 in Australia. Watson's instrument focused on capturing teachers' confidence to prepare and teach probability and statistics units, their beliefs about statistics, factors that are significant in teaching probability and statistics, and their confidence for teaching nine topics (chance language, equally likely outcomes, average, basic probability calculations, odds, median, graphical representation, data collection, and sampling). Even though these nine topics are specific to statistics, the topics do not capture the sophistication and progression across middle and high school, nor do they address critical parts of the curriculum such as statistical inference, association, and variability. Additionally, there is no published validity evidence or argument for score interpretations from the instrument.

The Self-Efficacy to Teach Statistics (SETS) Instruments

Development of the SETS instruments began in 2008, by a group of researchers who came began collaborating through a statistics education research cluster initiative, to address this critical need for a statistics-specific teaching efficacy instrument. There are two grade level–specific versions of the Self-Efficacy to Teach Statistics (SETS) instrument. The

Self-Efficacy to Teach Statistics—Middle Grades (SETS-MS) instrument measures mathematics teachers' self-efficacy for teaching statistics in grades 6–8, guided by the *GAISE Pre-K-12 Report* (Franklin et al., 2007) and the statistics topics in the grades 6–8 standards of the CCSSM. This instrument consists of 26 self-report, Likert-scale items (from 1 = "not at all confident" to 6 = "extremely confident") that ask the teacher to rate their "confidence to teach middle grade students the skills to" perform certain statistical tasks. The use of a 6-point response scale for the SETS instruments was chosen to align with the 6-point scale employed by Finney and Schraw's (2003) measures of college students' self-efficacy to learn statistics (Self-Efficacy to Learn Statistics scale; SELS) and to do statistics (Current Statistics Self-Efficacy scale; CSSE), regarding 14 concepts taught in a traditional college-level introductory statistics course. The SETS-MS has two subscales: "Level A—Reading the Data" (11 items) and "Level B—Reading Between the Data" (15 items). The analyses supporting the use of the 6-point scale and two subscales in scoring of SETS-MS responses are described subsequently.

The Self-Efficacy to Teach Statistics—High School (SETS-HS) instrument measures secondary (grades 9–12) mathematics teachers' self-efficacy for teaching statistics, guided by the *GAISE Pre-K-12 Report* and the two data analysis strands of the high school standards of the CCSSM. This instrument contains 44 self-report, Likert-scale items on a 6-point scale (from 1 = "not at all confident" to 6 = "extremely confident"). The first 26 items are identical to those in the SETS-MS instrument, while the last 18 are specific to the high school instrument. There are three subscales: "Level A—Reading the Data" (the first 11 items), "Level B—Reading Between the Data" (15 items), and "Level C—Reading Beyond the Data" (last 18 items). Subscale scores can be reported as either a sum of item responses or as mean item responses in order to report scores on the 6-point response scale. The analyses supporting the use of the 6-point scale and three subscales in scoring of SETS-HS responses are described subsequently.

Table 7.1 presents a sample of items from the SETS-MS (Levels A and B) and SETS-HS (Levels A, B, and C) instruments with the corresponding mathematics standards. Groth and Bargagliotti (2012) discuss the complementary nature of the GAISE and CCSSM, which is noteworthy as SETS-MS work began before the 2010 CCSSM release.

Additionally, the instrument developers wrote open-ended questions for the end of each subscale. These questions ask respondents to identify which items in each subscale were "easiest" and "hardest" to rate with the highest response category ("complete confidence") and the reasons for these ratings. These items were added during the development of the high school instrument to elicit what influences teachers' development of self-efficacy to teach statistics (Harrell-Williams, Sorto, Pierce, Lesser, & Murphy, 2014b). Responses to these items often provide information regarding

Table 7.1 Example SETS Items and Relevant Common Core Standards

Subscale	SETS-HS items	CCSSM Statistics Standards
"Reading the Data" (Level A)	Recognize that there will be natural variability between observations for individuals.	Develop understanding of statistical variability. Understand that a set of data has distribution with specific center/variation/shape.
	Create boxplots for summarizing distributions.	Summarize and describe distributions. Create boxplots/histograms. Summarize data numerically.
"Reading Between the Data" (Level B)	Recognize sampling variability in summary statistics, such as the sample mean and the sample proportion.	Use random sampling to draw inferences about a population. Use data from a random sample to draw inferences about a population with an unknown characteristic of interest.
	Use interquartile range, five-number summaries, and boxplots for comparing distributions.	Draw informal comparative inferences about two populations. Use graphs to estimate differences. Use numerical values to assess differences.
"Reading Beyond the Data" (Level C)	Identify the slope and y-intercept coefficients of a linear model and interpret them in the context of the data.	Interpret the slope (rate of change) and the intercept (constant term) of a linear model in the context of the data.
	Estimate percentages via the empirical rule (i.e., percentage of observations within 1, 2, or 3 standard deviations from the mean) using the mean and standard deviation of a data set which has an approximately bell-shaped distribution.	Use the mean and standard deviation of a data set to fit it to a normal distribution and to estimate population percentages. Recognize that there are data sets for which such a procedure is not appropriate.

Note: Originally printed in Harrell-Williams, Lovett, Lee et al. (2019), available electronically at https://journals.sagepub.com/doi/abs/10.1177/0734282917735151. Reprinted with permission of Sage Publications, Inc.

the teachers' comprehension of specific statistical topics on the SETS and their exposure to the material (e.g., "I'm not sure that I know what that is" or "That wasn't covered or emphasized in my statistics coursework").

While developing the SETS instruments, the researchers imagined several intended uses for the resulting individual responses and subscale scores. For mathematics and statistics educators teaching a single course or module within a course or conducting professional development activities, the responses and scores can provide feedback to the instructors about the impact of the course or session on teachers' self-efficacy to teach statistics. For teacher preparation program coordinators, the scores for teacher cohorts could provide information about the statistical topics needing more focus within the program or for a specific cohort. For preservice teachers, the instrument can be used as a benchmark tool as they progress through their teacher preparation program. For in-service teachers, the measure could help select professional development activities, in terms of which statistical topics should be highlighted in the professional development, or as a measure of change as individuals progress through professional development activities and the implementation of new lesson plans based on professional development experiences. Examples of such uses are cited in the section "Evidence of Instrument Use." However, the SETS instruments should not be used as an evaluation measure to allow entry into a teacher education program, to allow preservice teachers to progress to student teaching, or to signal readiness for graduation or teaching assignment.

Selecting a Validity Framework: Integrating Rasch Measurement Theory With Messick's Framework

Messick (1995) argued for a unified concept of validity that "integrates considerations of content, criteria, and consequences into a construct framework for empirically testing rational hypotheses about score meaning and theoretically relevant relationships" (p. 741). This approach differs greatly from the somewhat fragmented or haphazard approach to the collection and documentation of validity evidence that historically occurred. Wolfe and Smith (2007a, 2007b) illustrate both test development and data-based evidence for the six aspects from Messick's framework: content, substantive, structural, generalizability, external, and consequential. Hence, the works by Wolfe and Smith could be considered as a predominantly Rasch framework for validity evidence as this was seminal work within the Rasch measurement community.

While Messick's framework could have been implemented using classical test theory–based analyses, several elements of the Rasch framework add distinctive contributions to a validity argument. Rasch measurement philosophy is a strong measurement philosophy that establishes requirements of the items, and the instrument as a whole, to allow for invariance of item and person measures (Bond & Fox, 2007). That is, person and item estimates are sample independent.

Additionally, Rasch-specific analyses provide distinctive results related to the instrument calibration. An item response theory approach involves a statistically oriented approach of fitting multiple models (i.e., one-parameter and two-parameter logistic models) and choosing the estimated model that best fits the data. The opposite is true of Rasch analyses, which are driven by the desire to identify anomalies in person responses and item performance via Rasch model residuals to identify items on the instrument that need to be modified or removed to improve inferences from the instrument as a whole (Andrich, 2004a, 2004b; Boone, 2016). The analyses rely on person and item misfit statistics (e.g., infit and outfit) that are functions of the difference between actual responses and the expected responses from the appropriate Rasch model within the measurement context (e.g., the one-parameter Rasch model for unidimensional dichotomously scored instruments, the graded response model for unidimensional polytomously scored instruments, the MRCML model for multidimensional instruments, etc.). Additionally, Linacre (2004) developed a set of criteria for using Rasch model estimates and fit statistics to assess the performance of rating scales for polytomously scored instruments. This provides additional evidence about whether the chosen number of response options work in the manner in which they were intended by the instrument developers.

Lastly, with the development of a multidimensional Rasch model, remaining within a Rasch framework allowed for most analyses to be conducted using Rasch-based item and scale estimates. For instance, examination of factor structure using a multidimensional Rasch model allowed for a simpler item parameter estimation and factor structure evaluation process than the more roundabout, two-step implementation of a structural equation modeling approach. In the two-step approach, CFA is implemented in one software package (e.g., Mplus or Amos), followed by the estimation of Rasch item and person parameters as a separate step in a different software package (e.g., Winsteps or Conquest).

It should be noted that, while Rasch measurement theory presents opportunities for unique contributions to validity evidence, there are several challenges. More complex models, such as the multidimensional model, require larger sample sizes than classical test theory analyses. These sample sizes are often difficult to obtain for classroom-based research. Additionally, popular statistical software package developers were slow to include Rasch models, which meant that those interested in Rasch had to purchase Rasch-specific software, such as Winsteps or Conquest, or program their own models in SAS or R. More recently, this has become less of a concern as Rasch models have appeared in both SAS and R. PROC IRT was added to SAS in 2014, but it includes only unidimensional dichotomous and partial credit models. However, there are now several R packages such as eRM and TAM that run more complex models, as well as a freeware option, Jmetrik that runs both classical test theory and Rasch analyses.

Collecting a Body of Validity Evidence for the Use of SETS Scores

The SETS instruments' development and validation processes were heavily influenced by the Wolfe and Smith (2007a, 2007b) framework. The validity evidence documented during the SETS development and validation efforts is summarized in Table 7.2, which is followed by a discussion of each aspect with the Rasch-specific analyses described in some depth.

Content Evidence

Evidence for the content aspect of validity indicates how representative and relevant the items are to the intended content domain and assesses the technical quality of the items. Documentation of the development process contributes to the content validity evidence. Harrell et al. (2009), Sorto et al. (2010), and Harrell-Williams, Sorto, Pierce, Lesser, and Murphy (2014a) describe activities undertaken during the instrument development process for the SETS-MS. Development information regarding the SETS-HS is found in Harrell-Williams, Lovett, Lee et al. (2019). The SETS-MS item content was influenced by both the *GAISE Pre-K-12 Report* and a

Table 7.2 Validity Evidence Collected for SETS Instruments from 2008–2018

Validity Aspect (Messick, 1995)	Evidence From Development Phase	Data-Based Evidence
Content	Instrument purpose Item development process Expert reviews	Item technical quality analysis using Rasch residuals
Substantive	Item tryouts	Rating scale functioning analysis using Rasch item parameters Examination of Rasch item difficulty hierarchy
Structural	N/A	Confirmatory factor analysis in a multidimensional Rasch model framework
Generalizability	N/A	Reliability analysis Item calibration invariance
External	N/A	Correlations with other scales Group comparisons Examination of person-item maps to assess capacity to be responsive to change after intervention
Consequential	N/A	Discussion of intended and actual uses and interpretations of scores

Note: N/A indicates areas of development activities not discussed in Wolfe and Smith (2007b).

review of the 2007–2008 state standards from over 20 states that listed statistics content in either student or teacher expectations in documents available online. High school item content was influenced by the *GAISE Pre-K-12 Report* and the high school strands of the CCSSM. As part of an interactive poster session, attendees at the 2009 U.S, Conference on Teaching Statistics were used as experts to assess alignment of the SETS-MS items to the appropriate GAISE level (Sorto et al., 2010). Specifically, attendees were asked to match 10 SETS-MS items to both their corresponding GAISE expectation and six content knowledge assessment items, which were developed by SETS authors. While the SETS-MS and GAISE were nearly 100% aligned across the sets of items that the attendees chose to match, attendees' ability to match the SETS items and GAISE expectation to the knowledge items ranged from 50% to 90%, depending on the topic of the knowledge item.

Content validity evidence also includes examining Rasch-based residuals as indicators of item technical quality (Wolfe & Smith, 2007b). Specifically, the residuals measure the difference between observed item responses and those predicted by the Rasch model used in the analysis. The residuals are generally transformed to create two summary indicators of item performance. The unweighted mean-square statistic (i.e., outfit) indicates items for which unexpected responses were observed, but the responses are close to the respondent's ability. Weighted mean-square statistic (i.e., infit) indicates items for which unexpected responses are far from the respondent's ability. These statistics provide evidence of how well the each item functions and an item might need to be removed or revised in order to improve the measurement of the latent construct. While Linacre (2011) suggests the acceptable range of 0.50–1.50 for mean square fit statistics, Smith, Rush, Fallowfield, Velikova, and Sharpe (2008) argue for the use of a more conservative range of 0.75–1.30 in polytomous data.

A few of the SETS items were greater than either upper bound, indicating that responses to those items were not as close to the predicted responses as preferred (i.e., underfit). Analysis for the 26 SETS-MS items identified two (17 and 18, which address computing interquartile range and five-number summaries of data and using those values to compare distributions, respectively) with weighted and unweighted mean squares between the 1.30 cutoff suggested by Smith et al. (2008) and Linacre's 1.50 cutoff (Harrell-Williams et al., 2014a). The SETS-HS validation study of all 44 items found two items (12, distinguishing between statistical and deterministic questions, and 35, the identification and interpretation of regression coefficients) had weighted and unweighted mean square values between 1.30 and 1.50, and one (17, same item described previously) had values near 1.60 (Harrell-Williams, Lovett, Lee et al., 2019). These items were not removed because they were close to the acceptable range and only values greater than 2.0 suggest that an item distorts the measurement system (Linacre, 2011). However, they were flagged for possible wording changes in the instrument revision process described subsequently. Point-measure correlations also provide technical quality evidence by assessing how strongly items correlate with scores. Each SETS item had strong positive correlation (each $r > 0.57$) with its respective subscale score (Harrell-Williams et al., 2014a; Harrell-Williams, Lovett, Lee et al., 2019).

Substantive Evidence

The substantive aspect of validity is "the degree to which theoretical rationales relating to both item content and cognitive processing models adequately explain the observed consistencies among item responses" (Wolfe & Smith, 2007b, p. 207). Development-related substantive evidence for the SETS items focused on item tryouts with practicing mathematics teachers (Harrell et al., 2009; Harrell-Williams et al., 2014a), which led to wording revisions of some items.

Data-based substantive evidence included both the analysis of rating scale functioning and the confirmation of theoretical item hierarchy through Rasch analyses. Linacre (2004) discusses four *essential criteria* for validity purposes and four additional *helpful criteria* for evaluating the effectiveness of a chosen response scale (i.e., the 6-point scale of the SETS instruments). The *essential criteria* ensure that (1) each category was used by at least 10 respondents, (2) each category's responses followed a unimodal distribution, (3) the average person ability or trait estimate (in logits) increased with each category (i.e., those generally responding "little confidence" have lower self-efficacy estimates than those who generally respond "completely confident"), and (4) the average unweighted mean square fit statistic for each category (not item) was below 2.0. Three *helpful criteria* address item threshold parameters, the average value at which a respondent is more likely to use the next higher category. The first of the three criteria states that the thresholds should increase as categories increase, supporting the current ordering of the categories. The other two threshold-related criteria specify inter-threshold spacing, with the specific gap based on the number of points in the rating scale (*helpful criterion 2*) but no more than a maximum width of 5 logits distance (*helpful criterion 3*). The final *helpful* criterion addresses the coherence of the measure and the categories. "Measure implies categories" percentages report how frequently the expected rating from the Rasch model are actually observed in the data (i.e., how many times a person is expected to answer with a "3" on a scale of 1 to 4 actually answered with a "3"). Conversely, the "categories imply measure" percentages report how frequently an observed category matched the expected category from the Rasch model. Harrell-Williams et al. (2014a) and Harrell-Williams, Lovett, Lee et al. (2019) describe the detailed results of the rating scale analysis for the 6-point rating scale on the SETS instruments. For both instruments, all of the essential criteria were met. To examine the item difficulty hierarchy, Harrell-Williams, Sorto, Pierce, Lesser, and Murphy (2015) and Harrell-Williams, Lovett, Lee et al. (2019) present item-person maps (Wright maps) for the SETS-MS and SETS-HS, respectively. Specifically, the item difficulties for the higher developmental levels of the instrument (Levels B and C) are generally higher than those for Level A items.

Structural Evidence

Structural evidence conveys how well the scoring structure relates to the structure of the construct domain. Traditionally, Rasch-based dimensionality analyses focused on a confirmation of unidimensionality through principal component analysis of standardized Rasch-model residuals (Wolfe & Smith, 2007b) using Winsteps software (Linacre, 2011). With the development of the Multidimensional Random Coefficient Multinomial Logit Model (MRCMLM; Adams, Wilson, & Wang, 1997; Briggs & Wilson, 2003) and more modern Rasch software packages that implement the multidimensional Rasch model, such as Conquest (Adams, Wu, & Wilson, 2015) or the R package *tam* (Robitzsch, Kiefer, & Wu, 2018), Rasch-based confirmatory factor analysis (CFA) can be conducted. Multiple theory-based factor models are then estimated and model selection criteria, such as Akaike Information Criteria (AIC; Akaike, 1973) and Bayesian Information Criteria (BIC; Schwarz, 1978) are used, with smaller values of AIC and BIC indicating the superior model.

The levels of the GAISE Pre-K-12 framework that informed the development of the SETS instruments drove the choice of models compared in the CFAs. Specifically, evidence from Rasch-based CFA supported the use of two subscales for the SETS-MS over the use of one single statistics teaching efficacy score (Harrell-Williams et al., 2014a) and the use of three subscales for the SETS-HS over either a two-subscale scoring system (Level A and combined Levels B and C) or a single statistics efficacy score (Harrell-Williams, Lovett, Lee et al., 2019).

Generalizability Evidence

Generalizability refers to how well scoring properties maintain their meaning across measurement contexts, including different groups and settings. Reliability estimates are frequently provided as evidence of the generalizability aspect. Scores for the SETS-MS and SETS-HS instruments demonstrated acceptable reliability of separation in initial validation analyses with preservice mathematics teachers (Harrell-Williams et al., 2014a; Harrell-Williams, Lovett, Lee et al., 2019). Additionally, other analyses can be conducted that address generalizability, such as investigating measurement invariance across groups of examinees. Harrell-Williams, Lovett,

Pierce et al. (2017) found no evidence of differential item functioning on the SETS-MS across groups of preservice middle grades and preservice secondary mathematics teachers when overall teacher self-efficacy levels were taken into consideration. Additionally, no statistically significant difference was found in reliability estimates between the two groups.

External Evidence

The external aspect of validity indicates the degree that a measure is related to other measures of similar or dissimilar constructs. The presentation of correlations and assessment of theory-based group differences via hypothesis testing serve as external evidence. To this end, Harrell-Williams, Lovett, Lee et al. (2019) found that scores from each subscale of the SETS-HS, as well as the overall score, were highly correlated with preservice teachers' self-ratings of their ability to implement specific statistical skills taught in introductory-level college statistics classes as measured by the Current Statistics Self-Efficacy scale (Finney & Schraw, 2003). However, Lovett (2016) found that SETS-HS total scores were not highly correlated with total scores on the Levels of Conceptual Understanding of Statistics assessment (LOCUS; Jacobbe, 2015; Jacobbe, Case, Whitaker, & Foti, 2014) ($r = .22$), which is designed to measure the knowledge of the grades 6–12 statistics standards in the CCSSM. It should be noted that the few published studies investigating the relationship between mathematics content knowledge and teacher efficacy beliefs were conducted with elementary teachers and had inconsistent results. These studies have found moderately positive correlations (Swars, Smith, Smith, & Hart, 2009), weak positive correlations (Bates, Latham, & Kim, 2011; McCoy, 2011), or no significant correlation (Swars, Hart, Smith, Smith, & Tolar, 2007). These inconsistent results are partially due to the instruments being used to measure mathematical content knowledge and mathematics teaching efficacy. In terms of group comparisons, Harrell-Williams, Lovett, Pierce et al. (2017) found that two SETS-MS subscale scores were not statistically different when comparing preservice middle grades and preservice secondary mathematics teachers. While it would be worthwhile to do so, correlations with SETS-MS responses from middle grades teachers with the CSSE and LOCUS have not been collected as larger-scale data collection for the SETS-MS was completed prior to the release of the LOCUS and more recent work on the SETS instruments focused more on the High School version.

Additionally, Wolfe and Smith (2007b) discuss using person-item maps to measure an instrument's capacity to respond to change after an intervention. The person-item maps constructed for both the SETS-MS and SETS-HS using validation samples had some room for preservice teacher self-efficacy estimates to improve if a SETS instrument was used to measure change after pedagogical instruction or professional development experience, or change as one moves through a teacher preparation program. This 'space to grow' would indicate that there should not be a ceiling effect to limit the usefulness of either SETS instrument as a posttest measurement.

Consequential Evidence

Consequential evidence relates to "the value implications of score interpretation as a source for action" (Wolfe & Smith, 2007b, p. 224). The types of evidence they present support high-stakes decision-making about an individual, such as the standard process for setting cut scores, examination of raters, classification studies, and criterion selection. These types of analyses have not been conducted as their outcomes would not pertain to the intended uses of the SETS instruments. However, Messick (1995) included mention of social consequences, such as change in educational policies. The SETS developers imagined the use of the SETS instruments on a larger scale might be part of the discussion around mathematics teacher preparation policies, such as in conjunction with the ASA's *Statistical Education of Teachers* (Franklin et al., 2015). Indeed, Lovett and Lee (2017) used results from a cross-institutional study of 236 preservice teachers to make the case that the current efforts in teacher preparation are not preparing future teachers to meet the demands of the Common Core statistics standards. This article has the potential to impact teacher education policies and practices. For example, these results led to an NSF grant focused on developing modular materials to support statistics teacher education (DUE-1625713).

Recent Instrument Use

The SETS instruments have been used in at least three ways intended by the developers: (1) to evaluate the large-scale preparation of teachers to teach statistics (Lovett, 2016; Lovett & Lee, 2017); (2) for preservice and in-service teachers to evaluate their own growth in statistics teaching efficacy through engagement in a professional development or coursework (Harrell-Williams, Lovett, Koklu et al., 2017; Lee, Lovett, Peters, & Franklin, 2016); and (3) as an evaluation of teachers' growth in statistics teaching efficacy through professional development (Akoglu, 2018; Jones et al., 2017; Lee et al., 2016; Thrasher, Starling, Lovett, Doerr, & Lee, 2015). To evaluate teacher education programs on a large scale, the SETS-HS was used in conjunction with the LOCUS (Jacobbe et al., 2014) to examine the statistics teaching efficacy and statistical knowledge of preservice secondary mathematics teachers (Lovett, 2016; Lovett & Lee, 2017). The results of this implementation of the SETS-HS provided teacher education programs information about what statistical content preservice teachers felt the most and least confident to teach, as well as recommendations for statistical topics where teacher education programs should increase focus. The authors are unaware of individual mathematics teacher education programs using the SETS instruments to evaluate teacher cohorts to implement changes within their programs.

The SETS-MS and SETS-HS have been used in preservice education and professional development of in-service teachers to evaluate their own learning and for facilitators to evaluate the effectiveness of the professional development. Thrasher et al. (2015) incorporated the SETS-HS into a master's level statistics education course. Lee et al. (2016) discussed incorporating the SETS-HS into a massive open online class for professional development (TSDI MOOC, see http://go.ncsu.edu/tsdi). The SETS-HS was offered as a self-assessment in Units 1 and 5 of the course with participants being given total and subscale scores so they could self-monitor their confidence at the beginning and end of the course. The matched-pair scores were also used by Lee et al. (2016) and Akoglu (2018) to examine changes in confidence to teach statistics, showing significant increases in confidence to teach statistics in both studies. Jones et al. (2017) used the SETS-MS to evaluate the effectiveness of their hybrid professional development that combined an in-person professional development focused on increasing the teachers' statistical knowledge with an online professional development (the TSDI MOOC) and professional learning community focused on increasing the teachers' pedagogical statistical knowledge. Akoglu (2018) used the SETS-HS and other data sources from interviews, surveys, and discussion forums to also assess whether participation in a hybrid professional development increased confidence to teach statistics. In Akoglu's study, nine professional learning teams (63 participants) met weekly or biweekly while engaging in the TSDI MOOC. In each of these studies, researchers found that teachers' statistics teaching efficacy, as measured by the SETS, increased through the professional development experiences and identified specific aspects of the professional development that impacted some topics with which participants felt most confident.

Updating the SETS Instruments

Recent developments in K-12 statistics education have spurred the SETS authors to consider updating and improving the instruments. In 2009, at the time data-based validation activities started for the SETS-MS, the only statistics content knowledge assessment that was somewhat related to what preservice and practicing middle and secondary mathematics teachers would teach in grades 6–12 was the Comprehensive Assessment of Outcomes in a first Statistics course (CAOS; delMas, Garfield, Ooms, & Chance, 2007). This assessment was aligned to the GAISE college document and included formal statistical inference (e.g., t-tests for comparing two groups), whereas the CCSSM high school strands focus on informal, or randomization-based, inference through the use of simulations or bootstrapping. The authors documented validity evidence for its use with college students (delMas et al., 2007). Around the time when the SETS-HS instrument was released, the LOCUS project (Jacobbe, 2015; Jacobbe et al., 2014) was finalizing its assessments for measuring statistics knowledge covered by the CCSSM. Aligned with the *GAISE Pre-K-12 Report*, the LOCUS provides scores associated with the three GAISE levels, Beginning/Intermediate (Levels A and B for grades 6–12) and Intermediate/Advanced (Levels B and C for grades 10–12). Additionally, scores are available for each of the four steps in the statistical investigation process (Franklin et al., 2007): formulating statistical questions, collecting data, analyzing data, and interpreting results.

The second impetus for updating the SETS instruments was the TSDI MOOC, which focuses on teaching statistics using the four-phase statistical investigation process, introduces participants to both the SETS-HS and the LOCUS, and focuses on the use of real data and technology tools for engaging students in statistics (see Lee & Stangl, 2015 for description of course design). Hence, the SETS team began to investigate how to specifically align SETS with the LOCUS in terms of item coverage of the statistical investigation process and possible content-related scores in addition to the current scoring practice using GAISE levels. In support of the idea of possible content area scoring, a mixture Rasch analysis of the SETS-MS items found four items on the Level A subscale and three items on Level B subscale related to creating and using graphical displays of data distributions that contributed to differences between the latent classes of preservice middle and secondary teachers (Harrell-Williams et al., 2018a, 2018b).

Framing SETS Validity Arguments Through the Updated Standards for Testing

New developments in measurement theory also prompted a reflection on the SETS development and validation process. The latest revision of the *Standards for Educational and Psychological Testing* was jointly released in 2014 by the American Educational Research Association (AERA), the American Psychological Association (APA), and the National Council on Measurement in Education (NCME). The updated standards discuss five types of validity evidence, namely, evidence based on test content, response processes, internal structure, relations to other variables, and consequences of testing.

This chapter section integrates the updated testing standards framework into a validity argument for the SETS instruments using the previously collected evidence organized around Messick's framework. Similar to Bandalos' (2018) mapping of the updated standards to the historical types of validity, Table 7.3 organizes existing SETS validity evidence by

Table 7.3 SETS Validity Evidence (2008–2018) Organized by Types of Evidence (2014 Testing Standards) and Mapped to Work by Messick

Evidence Based On:	Validity Argument	Existing SETS Evidence	Mapping to Messick's Aspects of Evidence
Content	Items are appropriate and adequate for measuring the construct.	Item stem and wording relate to Bandura (2006) Alignment with GAISE and CCSSM during development process Expert reviews Item pilot study with teachers	Content
Response processes	Items tap into the intended cognitive processes.	Responses to open-ended items about reasons for ratings of specific items	Substantive (cognitive processing)
Internal structure	Items relate to other items as expected by theory.	CFA in a multidimensional Rasch model framework Reliability analysis Item calibration invariance Item difficulty hierarchy Item and rating scale technical quality analysis using Rasch residuals	Substantive (item difficulty hierarchy and rating scale analysis) Structural Generalizability
Relations to other variables	Scores from measure relate to other measures as expected by theory.	Correlations with CSSE and LOCUS Group comparisons	External
Consequences of testing	Intended consequences and unintended consequences are identified and explored.	Discussion of intended uses and interpretations of scores	Consequential

the types of validity evidence in the updated testing standards and maps these to the Messick frameworks. By explicitly stating the claim (second column) that each type of evidence supports, this table is a form of a validation argument for the SETS, using a combined Rasch measurement theory and updated Standards framework.

The mapping of existing SETS validity evidence onto the 2014 testing standards (AERA, APA, & NCME, 2014) was rather straightforward since Messick's framework is organized also by type of evidence. For content evidence, examples of possible evidence include a detailed definition of the construct, a table of specification, expert review, and identification of possible construct-irrelevant variance and construct-irrelevant underrepresentation. For the SETS instruments, particular attention was paid to the wording of the item stem as well as items to mirror the definition of self-efficacy and how to measure it based on Bandura's self-efficacy theory (Bandura, 2006). As mentioned, the statistical content contained in SETS items were grade-level appropriate based on alignment with the GAISE and CCSSM. In an attempt to reduce construct-irrelevant variance, a pilot study with teachers was used to identify potentially problematic wording during the development stage.

While response processes are subsumed under substantive evidence in Messick's framework, the updated standards list response processes as their own type of evidence. The intent of response process evidence is to verify that the appropriate cognitive processes are being employed as respondents interact with the items on the instrument. Possible evidence includes behavioral observations (such as eye-tracking or analysis of response times), think-aloud protocols or cognitive interviewing, expert–novice studies, and experimental studies manipulating item features. In our early work, we considered the pilot study data where the teachers provided feedback on item wording as partially addressing the response process. During development of the SETS-HS instrument, open-ended items were included, which asked teachers to identify one or two items within each subscale that they felt most efficacious or least efficacious about and to provide rationale for their ratings. Harrell-Williams et al. (2014b), Lovett (2016), and Lovett and Lee (2017) summarize findings about reasons preservice secondary teachers provide for their ratings of SETS-HS items. While some responses were about a lack of recent interaction with the material or not yet having encountered some concepts in their training, other responses indicated that their experiences as a tutor or student teacher informed their self-efficacy ratings. These responses are generally consistent with work on sources of self-efficacy (Bandura, 1977), where preservice teachers' mastery and vicarious experiences impact one's self-efficacy.

The internal structure evidence embodies how relationships among the items mirror those expected by theory. Such evidence includes correlations (items and subscales), internal consistency or reliability, factor analysis, generalizability theory, and exploration of differential item functioning. For the SETS instruments, the internal structure evidence category in the updated standards relates to multiple areas of Messick's framework: correlations and CFA in structural evidence, reliability and differential item functioning from generalizability, and item difficulty hierarchy maps from substantive evidence.

The evidence for relations to other variables in the updated standards is the same as Messick's external validity. Evidence includes correlations with other scales or variables, sensitivity and specificity analyses, and analyses of group differences. As mentioned in the external validity section, SETS-HS scores were correlated with scores for preservice secondary teachers' statistical knowledge and their ratings of self-efficacy to complete 14 statistical tasks normally taught in introductory statistics courses. As expected, the two self-efficacy scores were highly and positively correlated, but the SETS and LOCUS scores were positively and weakly correlated.

Similar to Messick's framework, the updated standards advocate for evidence based on the consequences of testing, including both intended and unintended consequences. As mentioned, the use of preservice teachers' SETS-HS responses as part of Lovett and Lee (2017) assessment of the current state of mathematics teacher preparation and the resulting award of research funding to develop curricular materials demonstrates one of the intended uses of the SETS instruments. The last column in Table 7.3 summarizes the connections between the updated standards and Messick's framework made in this section.

Future SETS Work

In September 2017, an instrument improvement process began. The SETS items were mapped onto the four steps of the statistical investigation process, creating a more comprehensive instrument blueprint, to identify gaps in existing item coverage. Focus groups were held with early instrument adopters to elicit feedback about their experience with using the SETS. Before and during the item writing and revision process, focus groups with preservice and in-service teachers, as well as mathematics and statistics educators, served to inform and refine item wording. A pilot study was being planned for February to March 2019 to initially evaluate the performance of the items on the revised instruments, and this would be followed by a large-scale data collection effort with preservice and in-service mathematics teachers.

Similar to how Lovett (2016) and Lovett and Lee (2017) built on the initial work of the SETS developers, other statistics and mathematics educators and teacher educators can contribute to the body of work for the newly revised SETS instruments. In concert with the SETS, a measure of content knowledge via the use of the LOCUS should be collected in order to assess correlation between teachers' statistics teaching self-efficacy and their grade level–appropriate statistics content knowledge. No widely used measure of mathematical knowledge for teaching specifically addresses the data analysis portions of the high school Common Core standards, inhibiting the assessment of the relationship between statistics teaching efficacy and pedagogical content knowledge. Comparisons across preservice teachers at different stages in their teacher preparation programs as well as across preservice and in-service teachers are also needed. Lastly, the use of the SETS instruments as measures of statistics teaching self-efficacy in relation to an expectancy-value theory model of instructor behavior and student achievement has not yet been investigated. For example, Batacki, Bolon, and Bond (2018) have proposed an instructor expectancy-value theory model, but are still in the early stages of item development to measure the other constructs in their model.

Implications for Assessment and the Construction of Validity Arguments

This chapter introduced the concept of statistics teaching efficacy, its importance in mathematics and statistics education research, and the potential value of the middle grades and high school versions of SETS instruments to researchers and teacher educators. This chapter also illustrates the use of a Rasch measurement theory-informed framework when gathering evidence for constructing a validity argument. Specifically, validity evidence collected over the last 10 years for both SETS instruments was summarized while introducing the Rasch-specific analyses that uniquely contributed to the validity argument for scores and uses of the SETS instruments. The authors hope that the presentation of this material might inspire others to consider Rasch measurement theory evidence when developing or substantially revising their mathematics- or statistics-related measures.

Acknowledgment

The early collaboration discussed in this chapter was supported in part by a National Science Foundation (NSF) grant (DUE-0618790) to the Consortium for the Advancement of Undergraduate Statistics Education (CAUSE).

References

Adams, R. J., Wilson, M., & Wang, W. C. (1997). The multidimensional random coefficients multinomial logit model. *Applied Psychological Measurement, 21*(1), 1–23. doi:10.1177/0146621697211001

Adams, R. J., Wu, M. L., & Wilson, M. R. (2015). *ACER ConQuest: Generalized item response modeling software (Version 4.0) [Computer software].* Melbourne: Australian Council for Educational Research.

Akaike, H. (1973). Information theory and an extension of the maximum likelihood principle. In B. N. Petrov & F. Csaki (Eds.), *Second international symposium on information theory* (pp. 267–281). Budapest: Akademiai Kiado.

Akoglu, K. (2018). *Blending online coursework and small learning communities to examine professional growth in teaching statistics: A phenomenological case study* (Unpublished doctoral dissertation). NC State University, Raleigh, NC.

Andrich, D. A. (2004a). Controversy and the Rasch model: A characteristic of incompatible paradigms? In E. V. Smith, Jr. & R. M. Smith (Eds.), *Introduction to Rasch measurement* (pp. 143–166). Maple Grove, MN: JAM Press.

Andrich, D. A. (2004b). Understanding resistance to the data-model relationship in Rasch's paradigm: A reflection for the next generation. In E. V. Smith, Jr. & R. M. Smith (Eds.), *Introduction to Rasch measurement* (pp. 167–200). Maple Grove, MN: JAM Press.

American Educational Research Association, American Psychological Association, National Council on Measurement in Education. (2014). *Standards for educational and psychological testing.* Washington, DC: AERA.

Ashton, P. T. (1985). Motivation and the teacher's sense of efficacy. In C. Ames & R. Ames (Eds.), *Research on motivation in education* (Vol. 2, pp. 141–171). Orlando, FL: Academic Press.

Bandalos, D. L. (2018). *Measurement theory and applications for the social sciences.* New York City, NY: Guilford Press.

Bandura, A. (1977). *Social learning theory.* Englewood Cliffs, NJ: Prentice-Hall.

Bandura, A. (2006). Guide for constructing self-efficacy scales. In F. Pajares & T. Urdan (Eds.), *Self-efficacy beliefs of adolescents* (pp. 307–337). Greenwich, CT: Information Age Publishing.

Bargagliotti, A., Anderson, C., Casey, S., Everson, M., Franklin, C., Gould, R., . . ., Watkins, A. (2014). Project-SET materials for the teaching and learning of sampling variability and regression. In K. Makar, B. de Sousa, & R. Gould (Eds.), *Proceedings of the ninth international conference on teaching statistics.* Voorburg, The Netherlands: International Statistical Institute. https://icots.info/9/proceedings/pdfs/ICOTS9_3E2_WATKINS.pdf

Batakci, L., Bolon, W., & Bond, M. (2018). A framework and survey for measuring instructors' motivational attitudes toward statistics. In M. A. Sorto, A. White, & L. Guyot (Eds.), *Proceedings of the tenth international conference on teaching statistics.* Voorburg, The Netherlands: International Statistical Institute. http://icots.info/10/proceedings/pdfs/ICOTS10_4J3.pdf

Bates, A. B., Latham, N., & Kim, J. A. (2011). Linking preservice teachers' mathematics self-efficacy and mathematics teaching efficacy to their mathematical performance. *School Science and Mathematics, 111*(7), 325–333.

Bond, T. G., & Fox, C. M. (2007). *Applying the Rasch model: Fundamental measurement in the human sciences* (2nd ed.). New York: Psychology Press.

Boone, W. J. (2016). Rasch analysis for instrument development: Why, when, and how? *CBE Life Sciences Education, 15*(4), rm4.

Briggs, D. C., & Wilson, M. (2003). An introduction to multidimensional measurement using Rasch models. *Journal of Applied Measurement, 4*(1), 87–100. doi:10.1007/s10763-013-9459-z

Calderhead, J. (1996). Teachers: Beliefs and knowledge. In D. C. Berliner & R. C. Calfee (Eds.), *Handbook of education psychology* (pp. 709–725). New York: Macmillan.

Certica Solutions. (2018). *Common Core State Standards adoption map*. Retrieved from http://statestandards.certicasolutions.com/common-core-state-adoption-map/

Cobb, G. W., & Moore, D. S. (1997). Mathematics, statistics, and teaching. *The American Mathematical Monthly, 104*(9), 801–823. doi:10.2307/2975286

delMas, R. (2004). A comparison of mathematical and statistical reasoning. In D. Ben-Zvi & J. Garfield (Eds.), *The challenge of developing statistical literacy, reasoning, and thinking* (pp. 79–95). The Netherlands: Kluwer Academic Publishers.

delMas, R., Garfield, J., Ooms, A., & Chance, B. (2007). Assessing students' conceptual understanding after a first course in statistics. *Statistics Education Research Journal, 6*(2), 28–58. https://iase-web.org/documents/SERJ/SERJ6(2)_delMas.pdf

Enochs, L. G., Smith, P. L., & Huinker, D. (2000). Establishing factorial validity of the mathematics teaching efficacy beliefs instrument. *School Science and Mathematics, 100*(4), 194–202. doi:10.1111/j.1949-8594.2000.tb17256.x

Finney, S. J., & Schraw, G. (2003). Self-efficacy beliefs in college statistics courses. *Contemporary Educational Psychology, 28*(2), 161–186. doi:http://dx.doi.org/10.1016/S0361-476X(02)00015-2

Fitzmaurice, O., Leavy, A., & Hannigan, A. (2014). Why is statistics perceived as difficult and can practice during training change perceptions? Insights from a prospective mathematics teacher. *Teaching Mathematics and its Applications, 33*(4), 230–248. doi:10.1093/teamat/hru010

Franklin, C., Bargagliotti, A. E., Case, C. A., Kader, G. D., Scheaffer, R. L., & Spangler, D. A. (2015). *Statistical education of teachers*. Alexandria, VA: American Statistical Association.

Franklin, C., Kader, G., Mewborn, D., Moreno, J., Peck, R., Perry, M., & Scheaffer, R. (2007). *Guidelines for Assessment and Instruction in Statistics Education (GAISE) report: A pre-K-12 curriculum framework*. Alexandria, VA: American Statistical Association.

Gould, R., Bargagliotti, A., & Johnson, T. (2017). An analysis of secondary teachers' reasoning with participatory sensing data. *Statistics Education Research Journal, 16*(2), 305–334. https://iase-web.org/documents/SERJ/SERJ16(2)_Gould.pdf

Groth, R., & Bargagliotti, A. E. (2012). GAISEing into the common core of statistics. *Mathematics Teaching in the Middle School, 18*(1), 38–45.

Harrell, L. M., Pierce, R. L., Sorto, M. A., Murphy, T. J., Lesser, L. M., & Enders, F. B. (2009). On the importance and measurement of pre-service teachers' efficacy to teach statistics: Results and lessons learned from the development and testing of a GAISE-based instrument. *Proceedings of the 2009 joint statistical meetings, section on statistical education* (pp. 3396–3403). Alexandria, VA: American Statistical Association.

Harrell-Williams, L. M., Lovett, J. N., Koklu, O., Lee, H. S., Sorto, M. A., Pierce, R. L., . . ., Franklin, C. (2017, May). *Using self-efficacy data to inform teacher preparation and professional development*. Breakout session at the 7th United States Conference on Teaching Statistics, State College, PA.

Leigh M. Harrell-Williams et al.

Harrell-Williams, L. M., Lovett, J. N., Lee., H. S., Pierce, R. L., Lesser, L. M., & Sorto, M. A. (2019). Validation of scores from the high school version of the Self-Efficacy to Teach Statistics (SETS-HS) instrument using pre-service mathematics teachers. *Journal of Psychoeducational Assessment, 37*(2), 194–208.

Harrell-Williams, L. M., Lovett, J. N., Pierce, -R. L., Sorto, M. A., Lee, H. S., & Lesser, L. M. (2017). The middle grades SETS instrument: Psychometric comparison of middle and high school pre-service mathematics teachers. In E. Galindo & J. Newton (Eds.), *Proceedings of the 39th annual meeting of the North American chapter of the international group for the psychology of mathematics education* (pp. 1064–1067). Indianapolis, IN: Hoosier Association of Mathematics Teacher Educators. Retrieved from www.pmena.org/pmenaproceedings/PMENA%2039%202017%20Proceedings.pdf

Harrell-Williams, L. M., Lovett, J. N., Sorto, M. A., Pierce, R. L., Lesser, L. M., & Murphy, T. J. (2018a, April). *Applying the Mixture Rasch Model to the Middle Grades Self-Efficacy to Teach Statistics (SETS-MS) instrument.* Paper presented at the International Objective Measurement Workshop (IOMW), New York City, NY.

Harrell-Williams, L. M., Lovett, J. N., Sorto, M. A., Pierce, R. L., Lesser, L. M., & Murphy, T. J. (2018b). Using the SETS Level a items to classify pre-service teachers' self-efficacy to teach statistics: An application of the Mixture Rasch Model. In M. A. Sorto (Ed.), *Proceedings of the tenth international conference on teaching statistics.* Kyoto, Japan: International Statistical Institute. https://iase-web.org/icots/10/proceedings/pdfs/ICOTS10_C220.pdf

Harrell-Williams, L. M., Sorto, M. A., Pierce, R. L., Lesser, L. M., & Murphy, T. J. (2014a). Validation of scores from a measure of teachers' self-efficacy to teach middle grades statistics. *Journal of Psychoeducational Assessment, 32*(1), 40–50. doi:10.1177/0734282913486256

Harrell-Williams, L. M., Sorto, M. A., Pierce, R. L., Lesser, L. M., & Murphy, T. J. (2015). Identifying statistical concepts associated with high and low levels of self-efficacy to teach statistics to middle grades. *Journal of Statistics Education, 23*(1), 1–20. doi:10.1080/10691898.2015.11889724 https://amstat.tandfonline.com/doi/abs/10.1080/10691898.2015.11889724

Jacobbe, T. (2015). *Developing K-12 teachers' understanding of statistics [Webinar].* Retrieved from www.causeweb.org/cause/webinar/teaching/2015-04/

Jacobbe, T., Case, C., Whitaker, D., & Foti, S. (2014). Establishing the validity of the LOCUS assessments through an evidence-centered design approach. In K. Makar, B. de Sousa, & R. Gould (Eds.), *Proceedings of the ninth international conference on teaching statistics.* Voorburg, The Netherlands: International Statistical Institute. https://icots.info/9/proceedings/pdfs/ICOTS9_7C2_JACOBBE.pdf

Jones, R. S., Lovett, J. N., & Google, A. (2017). Integrating face-to-face professional development and a MOOC-Ed to develop teachers' statistical knowledge for teaching. In E. Galindo & J. Newton (Eds.), *Proceedings of the 39th annual meeting of the North American chapter of the international group for the psychology of mathematics education* (p. 541). Indianapolis, IN: Hoosier Association of Mathematics Teacher Educators. www.pmena.org/pmenaproceedings/PMENA%2039%202017%20Proceedings.pdf

Lee, H. S., Lovett, J. N., Peters, S., & Franklin, C. (2016, April). *Teacher development in statistics education: A critical examination of how teachers' experiences impact their knowledge beliefs, and practices for teaching statistics.* Invited presentation at the annual National Council of Teachers of Mathematics Research Conference, San Francisco, CA.

Lee, H. S., & Stangl, D. (2015). Taking a chance in the classroom: Professional development MOOCs for teachers of statistics in K-12. *Chance, 28*(3), 56–63. doi:10.1080/09332480.2015.1099368

Linacre, J. M. (2004). Optimal rating scale category effectiveness. In E. V. Smith, Jr. & R. M. Smith (Eds.), *Introduction to Rasch measurement* (pp. 258–278). Maple Grove, MN: JAM Press.

Linacre, J. M. (2011). *Winsteps® (Version 3.73.0) [Computer software].* Beaverton, Oregon: Winsteps.com.

Love, A., & Kruger, A. C. (2005). Teacher beliefs and student achievement in urban schools serving African American students. *Journal of Educational Research, 99*(2), 87–98. doi:10.3200/JOER.99.2.87-98

Lovett, J. N. (2016). *The preparedness of preservice secondary mathematics teachers to teach statistics: A cross-institutional mixed methods study* (Unpublished doctoral dissertation). NC State University, Raleigh, NC.

Lovett, J. N., & Lee, H. S. (2017). New standards require teaching more statistics: Are preservice secondary mathematics teachers ready? *Journal of Teacher Education, 68*(3), 299–311. doi:10.1177/0022487117697918

McCoy, A. C. (2011). *Specialized mathematical content knowledge of preservice elementary teachers: The effect of mathematics teacher efficacy* (Unpublished doctoral dissertation). University of Missouri-Kansas City, Kansas City, MO.

McGee, J. R., & Wang, C. (2014). Validity-supported evidence of the self-efficacy for teaching mathematics instrument. *Journal of Psychoeducational Assessment, 32*(5), 390–403. doi:10.1177/0734282913516280

Messick, S. (1995). Validity of psychological assessment: Validation of inferences from person's responses and performances as scientific inquiry into score meaning. *American Psychologist, 50*(9), 741–749. doi:10.1002/j.2333-8504.1994.tb01618.x

National Council of Teachers of Mathematics. (1989). *Curriculum and evaluation standards for school mathematics.* Reston, VA: Author.

National Council of Teachers of Mathematics. (2000). *Principles and standards for school mathematics.* Reston, VA: Author.

National Governors Association Center for Best Practices and Council of Chief State School Officers. (2010). *Common core state standards for mathematics.* Washington, DC: Authors.

Pajares, F. (1997). Current directions in self-efficacy research. In M. Maehr & P. R. Pintrich (Eds.), *Advances in motivation and achievement* (Vol. 1, pp. 1–49). Greenwich, CT: JAI Press.

Peters, S. A., Watkins, J. D., & Bennett, V. M. (2014). Middle and high school teachers' transformative learning of center. In K. Makar, B. de Sousa, & R. Gould (Eds.), *Proceedings of the ninth international conference on teaching statistics.* Voorburg, The Netherlands: International Statistical Institute. https://icots.info/9/proceedings/pdfs/ICOTS9_C151_PETERS.pdf

Philipp, R. (2007). Mathematics teachers' beliefs and affect. In F. K. Lester (Ed.), *Second handbook of research on mathematics teaching and learning* (pp. 257–315). Charlotte, NC: Information Age Publishing.

Pierce, R., & Chick, H. L. (2011). Teachers' beliefs about statistics education. In C. Batanero, G. Burrill, & C. Reading (Eds.), *Teaching statistics in school mathematics: Challenges for teaching and teacher education* (pp. 151–162). The Netherlands: Springer.

Robitzsch, A., Kiefer, T., & Wu, M. (2018). *TAM: Test analysis modules. (R package Version 2.9–35) [Computer software].* Retrieved from https://CRAN.R-project.org/package=TAM

Rossman, A., Medina, E., & Chance, B. (2006). A post-calculus introduction to statistics for future secondary teachers. In A. Rossman & B. Chance (Eds.), *Proceedings of the seventh international conference on teaching statistics.* Salvador, Brazil: International Statistics Institute. https://iase-web.org/documents/papers/icots7/2E2_ROSS.pdf

Schwarz, G. (1978). Estimating the dimension of a model. *Annals of Statistics,* 6(2), 461–464. doi:10.1214/aos/1176344136

Smith, A. B., Rush, R., Fallowfield, L. J., Velikova, G., & Sharpe, M. (2008). Rasch fit statistics and sample size considerations for polytomous data. *BMC Medical Research Methodology,* 8(33), 1–11. doi:10.1186/1471-2288-8-33 https://bmcmedresmethodol.biomedcentral.com/articles/10.1186/1471-2288-8-33

Sorto, M. A., Harrell, L. M., Pierce, R. L., Murphy, T. J., Enders, F. B., & Lesser, L. M. (2010). Experts' perceptions in linking GAISE guidelines to the self-efficacy to teach statistics instrument. In *Proceedings of the 2010 joint statistical meetings, section on statistical education* (pp. 4289–4294). Alexandria, VA: American Statistical Association.

Swars, S. L., Hart, L. C., Smith, S. Z., Smith, M. E., & Tolar, T. (2007). A longitudinal study of elementary pre-service teachers' mathematics beliefs and content knowledge. *School Science and Mathematics,* 107(8), 325–335.

Swars, S. L., Smith, S. Z., Smith, M. E., & Hart, L. C. (2009). A longitudinal study of effects of a developmental teacher preparation program on elementary prospective teachers' mathematics beliefs. *Journal of Mathematics Teacher Education,* 12(1), 47–66. https://doi.org/10.1007/s10857-008-9092-x

Thrasher, E., Starling, T., Lovett, J. N., Doerr, H. M., & Lee, H. S. (2015). The influence of a graduate course on teachers' self-efficacy to teach statistics. In T. G. Bartell, K. N. Bieda, R. T. Putnam, K. Bradfield, & H. Dominguez (Eds.), *Proceedings of the thirty-seventh annual meeting of the North American chapter of the international group for the psychology of mathematics education* (pp. 447–454). East Lansing, MI: Michigan State University. www.pmena.org/pmenaproceedings/PMENA%2037%202015%20Proceedings.pdf

Watson, J. M. (2001). Profiling teachers' competence and confidence to teach particular mathematics topics: The case of chance and data. *Journal of Mathematics Teacher Education,* 4(4), 305–337. doi:10.1023/A:1013383110860

Wilkins, J. L. (2008). The relationship among elementary teachers' content knowledge, attitudes, beliefs, and practices. *Journal of Mathematics Teacher Education,* 11(2), 139–164. doi:10.1007/s10857-007-9068-2

Williams-Harrell, L. M., Sorto, M. A., Pierce, R. L., Lesser, L. M., & Murphy, T. J. (2014b). Using the SETS instruments to investigate sources of variation in levels of pre-service teacher efficacy to teach statistics. In K. Makar, B. de Sousa, &

R. Gould (Eds.), *Proceedings of the ninth international conference on teaching statistics*. Voorburg, The Netherlands: International Statistical Institute. https://icots.info/9/proceedings/pdfs/ICOTS9_C270_HARRELLWILLIAMS.pdf

Wolfe, E. W., & Smith, E. V. (2007a). Instrument development tools and activities for measure validation using Rasch models: Part I—Instrument development tools. *Journal of Applied Measurement, 8*(1), 97–123.

Wolfe, E. W., & Smith, E. V. (2007b). Instrument development tools and activities for measure validation using Rasch models: Part II—Validation activities. *Journal of Applied Measurement, 8*(2), 204–234.

Zee, M., & Koomen, H. Y. (2016). Teacher self-efficacy and its effects on classroom processes, student academic adjustment, and teacher well-being: A synthesis of 40 years of research. *Review of Educational Research, 86*(4), 981–1015. doi:10.3102/0034654315626801

8 Measurement and Validity in the Context of Mathematics Coaches

Kristin E. Harbour, Stefanie D. Livers, and Margret A. Hjalmarson

Over the last 20 years, a growing number of states and school districts have employed mathematics coaches in response to needs such as providing school-based, ongoing professional development, analyzing assessment data, and navigating curriculum changes successfully (McGatha & Rigelman, 2017a). The need to provide ongoing, school-based support for in-service teachers is well acknowledged, and in some cases has been filled by creating positions in schools and districts that focus on improving mathematics teaching and learning (e.g., Darling-Hammond, Hyler, & Gardner, 2017; Woulfin & Rigby, 2017). The term *mathematics coach* is often used when referring to these in-school support positions. We have adapted the term *mathematics coach* in this chapter following Campbell and Malkus' (2013) description as follows:

> The role of the mathematics specialist or coach is to support the improvement of mathematics teaching and learning in schools by targeting teachers' understanding and action. The assumptions are that the specialist or coach is a knowledgeable colleague who has pedagogical expertise and an understanding of mathematics and of how students learn and that this person is qualified and capable of serving as an on-site resource and leader for teachers, providing school-based and content-specific professional development.
>
> (pp. 213–214)

With the increased presence, training, and research on mathematics coaches (e.g., Fennell, 2017; McGatha, Davis, & Stokes-Levine, 2017), a need for measures that produce interpretations with strong validity evidence is at the forefront. The evolving job description and duties, which vary by school, district, and state; the evolving understanding in the field; and the evolving understanding of the knowledge and skills needed all create challenges and opportunities for measurement development. Without instruments that accurately measure specific constructs pertaining to mathematics coaching, state, district, and school leaders are left to make policy decisions on the use of mathematics coaches without empirical evidence to support the claims regarding the coaches' work in schools. The lack of measures with valid outcomes limits the advancement of the research pertaining to mathematics coaches in relation to large-scale, multistate, multidistrict comparisons. With the majority of evidence being qualitative in nature, the scaling up of research on mathematics coaching is limited and the generalizability of the results is prohibited. In an effort to advance the work done by mathematics coaches and the research with mathematics coaches, we will first discuss the participants in mathematics coaching research and the existing frameworks and instruments, then describe the needed instruments, and conclude with challenges and considerations of instrumentation.

Participants Involved in Mathematics Coaching Research

In this section, we explore the complexity of the roles of mathematics coaches (sometimes referred to as mathematics specialists, teacher leaders, or interventionists) which may lead to difficulty in identifying and designing appropriate measures to capture the nuanced work of these school-based mathematics instructional leaders. An overarching distinction is often made when defining the terms *coach*, *specialist*, *leader*, or *interventionist* when viewing the primary responsibility of this individual: The individual is primarily tasked with supporting teachers (coach or teacher leaders) versus primarily tasked with supporting students (specialists or interventionist); however, this distinction is not always as clear. Different schools or districts may use the same term to identify different roles depending on state regulations, certification programs, and varying responsibilities of the individual within their school or district. The Association for Mathematics Teacher Educators (AMTE) adopts a broad definition as well: "Elementary mathematics specialists are teachers, teacher leaders, or coaches who are responsible for supporting effective mathematics instruction and student learning at the classroom, school, district, or state levels" (2013, p. 1). The work of a mathematics coach is not the same as the work of a mathematics teacher yet "elementary mathematics specialist" may be the title used for both roles depending on the context. So, for example, if a survey asks "Does your school have a mathematics specialist?", two different principals might answer yes even though the mathematics specialists in one school is primarily responsible for teaching children and the mathematics specialist in the other school might be responsible for supporting teacher professional development.

Defining the Various Titles and Roles

As early as 1984, John Dossey called for elementary mathematics specialists in an article called "Elementary School Mathematics Specialists: Where Are They?" We continue to have the same questions about where they are, but we are also asking what mathematics coaches, specialists, leaders, and interventionists do. Our discussion in this section focuses on the nature of the work and the diversity of tasks that these instructional leaders might engage in. The term *mathematics specialist* is sometimes used to define a teacher who specializes in teaching mathematics (e.g., an elementary teacher who may teach only mathematics or may teach mathematics and science; Markworth, Brobst, Ohana, & Parker, 2016). McGatha and Rigelman distinguish between *mathematics intervention specialists* who work with selected students (e.g., in small group for extra support) and *mathematics coaches* who are "professionals who primarily work with teachers" (2017b, p. xiv). A *mathematics coach* may work with individual teachers or may facilitate the work of groups of teachers (Barlow, Burroughs, Harmon, Sutton, & Yopp, 2014; Elliott et al., 2009; Lesseig et al., 2016). Figure 8.1 shows how we have defined the different titles and where the emphasis of their work lies in schools.

A series of standards documents[1] from AMTE (2013) and the National Council of Teachers of Mathematics (NCTM, 2012) have emerged over the last 10 years that list the knowledge and skills that mathematics coaches should have. These documents focus primarily on the mathematics content knowledge and pedagogical content knowledge (e.g., lesson planning, understanding technology for mathematics learning) but also include some description of the leadership knowledge and skills for mathematics coaches and specialists. However, these lists are wide-ranging and include duties that may extend from the district to the school to the classroom, from work with parents, administrators, teachers, or students. While there are a wide range of activities in which a mathematics coach or specialist might engage, not all coaches or specialists engage in all activities all the time (Campbell & Malkus, 2013, 2011). As a result, an ongoing challenge

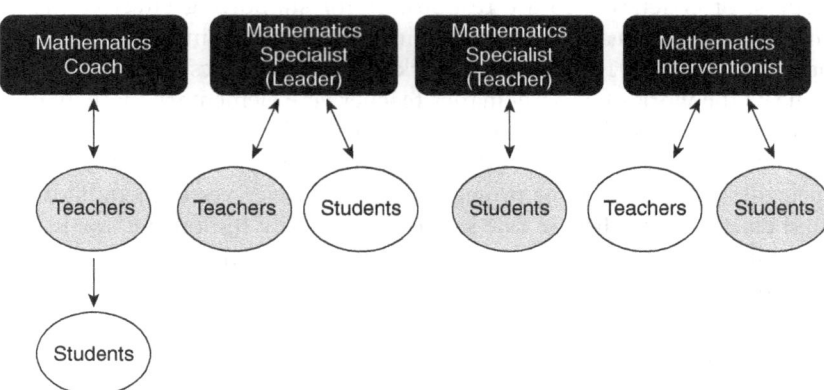

Figure 8.1 Job titles and purposes for mathematics instructional leaders. Shaded bubbles indicate the primary job purpose of the coach, specialist, and interventionist.

of research and professional development related to mathematics coaches and specialists is the range of terms used to describe their roles in schools and their work with teachers. Although each role (e.g., coach, specialist, interventionist) is critical in advancing the teaching and learning of mathematics, our focus is on mathematics coaches and their primary role of supporting teachers.

Mathematics Coaches

There are multiple models that are adopted by schools and districts to guide the work of mathematics coaches that can be divided into different levels (see Figure 8.2). At the first level is coaching work with individual teachers or pairs of teachers (e.g., Barlow, Burroughs, Harmon, Sutton, & Yopp, 2014). This work often follows a timeline of planning, co-teaching/observation, and debriefing with the teacher and the coach ideally working collaboratively (West & Cameron, 2013; West & Staub, 2003). At the next level, the coach might be responsible for working with groups of teachers (e.g., facilitating a professional learning community or a grade-level team; Elliott et al., 2009; Lesseig et al., 2016). In this work, the mathematics coach might be supporting team-level planning and analysis of students' learning. The mathematics coach might work

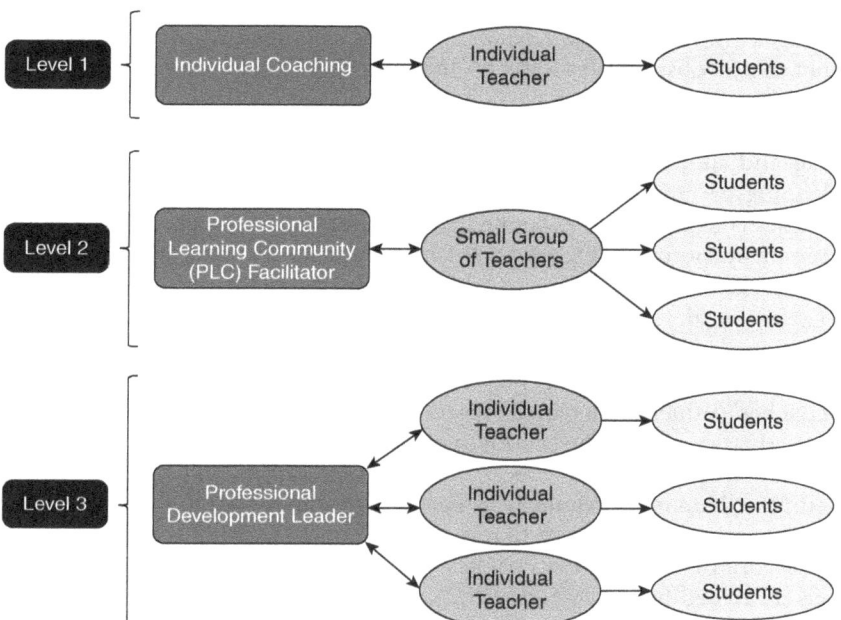

Figure 8.2 Levels of interaction within the context of mathematics coaching.

with the team on particular teaching practices or common initiatives. At the third level, the mathematics coach might be focused on school-level initiatives, which could include the analysis of assessment data, planning school-level professional development, or other work with administrators or other staff to carry out activities focused on teaching and learning (Campbell, Ellington, Haver, & Inge, 2013; Felux & Snowdy, 2006; McGatha & Rigelman, 2017a).

A common benefit of the work of the mathematics coach is that recommendations for teacher professional development commonly advocate for ongoing, school-based initiatives and support for teacher learning (Guskey, 2000; Joyce & Showers, 2002). The mathematics coach is by definition positioned to both have knowledge of mathematics teaching and learning and be the ongoing supporter and facilitator at the school level for teacher professional development. For this reason, the role of mathematics coach can be helpful in bridging research to practice. From a research perspective, the role of the mathematics coach cannot be ignored when analyzing data about the impact of a professional development program on teacher knowledge, beliefs, and practice. Just as students' learning should be analyzed by including the teacher in that analysis, teachers' learning should be analyzed with the mathematics coach in mind since they can play a key role in the design and implementation of professional development for teachers.

Within research projects, the mathematics coach could play a role as the facilitator of professional development at the school level or across the district (Borko, Jacobs, Koellner, & Swackhammer, 2015; Borko, Koellner, & Jacobs, 2014; Cobb & Jackson, 2011). The project then could have multiple layers of participants, including both students, teachers, and mathematics coaches. The project might also need to provide training and support for the mathematics coach if they are responsible for leading or facilitating professional development initiatives. The mathematics coach could also be the focus of the study if the project is examining professional development for mathematics coach specifically (Cobb, Henrick, & Munter, 2011; Cobb & Jackson, 2015; Elliott et al., 2009; Lesseig et al., 2016). Both of these scenarios create measurement needs that are similar to understanding the influence of teachers on students' learning. At this level, there needs to be documentation of mathematics coaches' influence on teacher learning and development or their influence on other aspects of the school/district initiative.

As states begin to have more mathematics coaches and schools begin to employ them more regularly as part of the school staff, it raises questions about policy and practice that require means of gathering evidence about the work of mathematics coaches. In order to compare the different models described here (see Figure 8.2), instruments will need to capture that variation in the role of the mathematics coach.

Existing Frameworks and Instruments

Within the current body of literature on mathematics coaches, there are frameworks and instruments used to study the roles, work, and impacts of mathematics coaches. Many of these frameworks and instruments are designed to study mathematics teaching and learning, not necessarily situated in the context of mathematics coach work. Others are more general in nature and stem from the leadership lens in terms of teacher support. As the increasingly urgent call for mathematics coaches has grown, frameworks and instruments that are more specific to this role and responsibility have emerged to examine the mathematics coaches balance of the three areas of knowledge: mathematics pedagogical content knowledge, content knowledge, and leadership knowledge (Bitto, 2015).

Existing Frameworks

There are numerous frameworks that capture the dynamics of the work and influences of mathematics coaches. These frameworks often provide the elements of inquiry for research, as mathematics coaching is a valid form of professional development to foster ongoing, sustained support for teachers (Guskey, 2000; Joyce & Showers, 2002). Other frameworks used for mathematics professional development of teachers are specifically related to the practices of mathematics; these allude to the roles and responsibilities, but do not intentionally identify a mathematics coach or specialist (Loucks-Horsey, Stiles, Mundry, Love, & Hewson, 2010; Smith, 2001).

Research frameworks sometimes seek to define the process of the work of mathematics coaches, including cognitive apprenticeship (Collins, Brown, & Newman, 1987; Harbour, Adelson, Karp, & Pittard, 2018) and social network analysis (Hopkins, Ozimek, & Sweet, 2017). Some frameworks focus on shifting instructional practices through the coaching conversation (Costa & Garmston, 1994; McGatha & Bay-Williams, 2013). While other frameworks fall into a model for professional development and the complexity for instructional change (Fullan & Stiegelbauer, 1991), others are situated within educational leadership structures and models (Bitto, 2015; Fennell, Kobett, & Wray, 2017). The National Council of Supervisors of Mathematics (NCSM) created the PRIME Leadership Framework as a structure for leaders in mathematics education (2008). As many coaches are involved within mathematics professional learning communities (PLCs; DuFour, 2004), occasionally this is the chosen framework to position the research within a community of practice model (Dean & McClain, 2006; Gibbons, 2017; Sowder, 2007). Additionally, a situative perspective on cognitive and learning is used to frame coaching within communities of practice (Greeno, Moore, & Smith, 1993; Lave & Wenger, 1991).

The frameworks described here, although not exhaustive, depict accurate descriptions for the relationship and behaviors that mathematics coaches possess. This was the intent of these frameworks; however, in the absence of measurement tools, the frameworks are often used in lieu of an instrument. The limitations of frameworks for research include: (1) lack of criteria for assessing the work and influence of the mathematics coach and (2) absence of specific behaviors or models tied to student achievement or instructional change.

Existing Tools/Instruments

The field is not without existing instruments for empirical research studies regarding the work of mathematics coaches. As we consider the three areas of expertise (content, pedagogy, and leadership) outlined in the *Standards for Elementary Mathematics Specialists* (AMTE, 2013), often each area is studied with instruments specific to each individual area. Content assessments for mathematics are available such as the Learning Math for Teaching (LMT; Hill, Sleep, Lewis, & Ball, 2007) and the *Diagnostic* Teacher Assessment in Mathematics and Science (DTAMS; Center for Research in Mathematics and Science Teacher Development, 2008) to measure mathematical content knowledge of mathematics coaches.

Classroom observation instruments are often used to examine pedagogical practices that have been identified in research as ways to advance the teaching and learning of mathematics (Bostic, Lesseig, Sherman, & Boston, in press; Boston, Bostic, Lesseig & Sherman, 2015). The observation instruments range in depth and breadth of capturing effective pedagogy of mathematics coaches. The key in choosing an observation instrument is to define the practices to be examined (Bostic et al., in press; Boston et al., 2015). The Mathematical Quality of Instruction (MQI; Hill, 2010) and the Mathematics Scan (M-SCAN; Berry, Rimm-Kaufman, Ottmar, Walkowiak, & Merritt, 2010) are used on video-recorded mathematics lessons. Instruments used for live observations or video-recorded lessons include the Classroom Observation Scales (Secada & Byrd, 1993), Mathematics Classroom Observation Protocol for Practices (MCOP[2]; Gleason, Livers, & Zelkowski, 2017), Oregon Collaborative for Excellence in the Preparation of Teachers' (OCEPT) Classroom Observation Protocol (OTOP; Wainwright, Flick, & Morrell, 2003), and the Revised SMPs Look-for Protocol (Bostic, Matney, & Sondergeld, 2019). These classroom observation instruments provide a range of varying criteria and scales for capturing different elements of mathematics teaching and learning. An additional "observation" of assessment of teaching tools exists that does not include a formal observation. The Instructional Quality Assessment (IQA) is a comprehensive instrument and uses both artifacts and observation to capture the mathematics pedagogy (Boston & Wolf, 2006). The IQA is focused on the selection and implementation of mathematics tasks.

Instruments have also been designed to capture leadership qualities and behaviors of mathematics coaches. These instruments include surveys (e.g., Elementary Mathematics Specialists & Teacher Leader Project [ems&tl], 2012; MIST, Cobb et al., 2011; NAEP, 2019) and interview tools (e.g., MIST, Cobb et al., 2011) that document the duties and responsibilities of mathematics coaches. There are survey instruments that highlight the relationship within a coaching conversation (Examining Mathematics Coaching's [EMC] Coaching Reflection and Impact Survey, Yopp, 2008a; Yopp, Rose, & Meade, 2008; TRU Math Conversation Guide, Baldinger & Louie, 2014), and instruments to assess coaching skills (EMC's Coaching Skills Inventory, Yopp, 2008b). In addition, there are daily logs that seek to capture specifics regarding time spent on tasks (Campbell & Griffin, 2017; Campbell & Malkus, 2010; Whitenack & Ellington, 2007). Many mathematics coaches find themselves leading professional learning communities. The Professional Learning Community Assessment (PLCA; Hipp & Huffman, 2003) is one tool that assesses the progress of the PLC. Killion, Harrison, Bryan, and Clifton (2012) provide several PLC tools that incorporate the role of a coach in the process. Mathematics-specific PLC tools for lesson design, teaching analysis, common assessment development, and post-lesson reflection can be found in Kanold and Larson (2012). These tools are more qualitative in nature.

There are some comprehensive instruments and scales based on conceptual frameworks. The Concerns-Based Adoption Model (CBAM; Hall, Wallace, & Dossett, 1973) examines three components within the process of supporting changes in instruction: Stages of Concerns (SoC), Levels of Use (LoU), and Innovation Configurations (IC). CBAM offers instruments to assess the three components for evaluation of coaching support of professional change. The TRU Mathematics Dimensions (Schoenfeld, Floden, & the Algebra Teaching Study and Mathematics Assessment Project, 2014) provides an analytic framework for mathematics classroom activities and a scoring rubric designed to measure the five dimensions of productive mathematics classrooms and is not meant for scoring classroom teaching and learning.

There remains a limited number of instruments specific to the work of mathematics coaches, with most developed from research grant projects. The Examining Mathematics Coaching project resulted in five instruments: EMC Coaching Skills Inventory, EMC Coach Reflection and Impact Survey, EMC Teacher Needs Inventory, EMC Teacher Reflection and Impact Survey, and EMC Teacher Survey (Yopp, 2008a, 2008b; Yopp et al., 2008). These instruments seek to identify relationships between the mathematics coach and the teacher; the skills, knowledge, and pedagogy of the mathematics coach; and the background of the mathematics coach. The Elementary Mathematics Specialists & Teacher Leader Project (ems&tl, 2012) developed a log to document the day-to-day activities of mathematics coaches, administered once a year for an overview of the mathematics coaches' work. The Middle-School Mathematics and the Institutional Setting of Teaching project (MIST; Cobb et al., 2011) at Vanderbilt University developed a survey that is designed to capture mathematics coaches' demographic information and actions with teachers, principals, and other school and district personnel; it also documents involvement in professional development. A second instrument from the MIST project was designed as an interview tool to understand participants' vision of high-quality instruction for both coaches and teachers (Munter, 2014). The only instrument or measure to date that provides researchers with large-scale, nationally representative data regarding mathematics coaches is the National Assessment of Educational Progress (NAEP, 2019); this survey is completed by school administrators (i.e., principal or vice principal) which may cause reliability concerns. See Table 8.1 for a comparison of instruments specific to mathematics coaching.

Instruments Needed for Research Involving Mathematics Coaches

The goals of mathematics coaches are to influence teacher practices and beliefs to in turn improve student learning, understanding, and performance (AMTE, 2013; Campbell & Malkus, 2013; McGatha & Rigelman, 2017b; Lesseig et al., 2016). To achieve the goals of positively influencing teachers and students, fine-grained details about knowledge, skills, practices, facilitation, and responsibilities are needed in order achieve the overarching goals typically attributed to the work of mathematics coaches. With the current measures available, however, we are left with fragmented evidence to make the connections among these goals. To advance the research and work of mathematics coaches, instruments are needed to measure: (1) leadership knowledge, practices, and skills; (2) facilitation and communication; and (3) roles, responsibilities, and impact. Table 8.2 provides a sample of overarching hypothetical research questions and instruments needed gather evidence to answer them. In the sections that follow, we expand on the ideas presented in Table 8.2 with examples of needed instruments in each category and a rationale for their development. Assessments could vary in terms of who is completing them (e.g., a coach, a teacher, a researcher, a principal) and whether the protocol is observing practice versus reporting on practice. The choice of who needs to complete an instrument and the context for the data collection will depend on the question being asked about coaching.

Instruments to Measure Leadership Knowledge, Practices, and Skills

Numerous instruments are available to measure content and pedagogical content knowledge needed for coaching (e.g., Yopp, 2008a, 2008b;

Table 8.1 Existing Instruments Focused on Mathematics Coaches

Instrument	Author	Focus	Target	Type	Items	Development	Reliability
Examining Mathematics Coaching (EMC) Coaching Skills Inventory (CSI)	Yopp (2008b)	Various coaching skills	Mathematics coaches	Survey	Relationships, skills, content, pedagogy, and background	Factor analysis of 24 survey items	> .80
Examining Mathematics Coaching (EMC) Coaching Reflection & Impact Survey (CRIS)	Yopp (2008a) Yopp et al. (2008)	Monitoring and logging coaching interactions; perspectives of impact	Mathematics coaches	Survey	Quantity, quality, and duration of coaching sessions; topics discussed	Factor analysis of 17 coaching-reflection items and 13 coaching-impact items	> .89
Middle School Mathematics and the Institutional Setting of Teaching (MIST)	Cobb et al. (2011)	Coaches' networks and roles	Mathematics coaches	Interview	Coaching practices, relationships, legitimacy	Three rubrics; coded 600 interviews	Not applicable
Middle School Mathematics and the Institutional Setting of Teaching (MIST)	Cobb et al. (2011)	Coaches' work and training	Mathematics coaches	Survey	Work with teachers, principals, professional development	Items came from other sources, and some items field tested with cognitive interviews	> .79
The Instructional Specialist Activity Manager (ISAM)	Campbell and Malkus (2010)	Daily activities	Mathematics coaches	Daily logs	Daily time spent on activities	Menu items with categories and subcategories	Not found
Mathematics Specialists-School District Model Survey	ems&ctl (2012)	Coaches' data at the district level	Mathematics coaches	Survey	Duties, roles, responsibilities	Menu of items	Not found
National Assessment of Educational Progress (NAEP)	NAEP (2019)	Coaches' data at the school level	Administrators of mathematics coaches	Survey	Presence of mathematics coaches, roles and responsibilities	Likert scale of time engaged in various roles and responsibilities	Not found

Note: Reliability reported for the overall instrument. Some instruments lacking readily available information on reliability.

Table 8.2 Instruments Needed for Potential Research Questions About Mathematics Coaches

Hypothetical Research Questions	Instruments Needed
Who are math coaches?	Set of survey items about the roles of mathematics coaches: • Different survey items for the coach, teacher, administrator, and district coordinator to address different perspectives and definitions • Capture different models (e.g., school based, district level, part time, full time)
What work do mathematics coaches engage in?	Set of survey items about the responsibilities of mathematics coaches: • Different survey items for the coach, teacher, administrator, and district coordinator to capture the similarities and differences among stakeholders • Different survey items to address the different levels of measurement to capture the vision of their responsibilities (administrator and/or coach reported) versus day-to-day responsibilities (coach reported) • Different administration time frames to capture change at different time scales (e.g., week to week, different responsibilities over a school year, year-to-year variation)
How do math coaches facilitate teachers' learning?	Observation instruments, teacher reflection protocols, coach reflection protocols (e.g., survey items, interview protocols): • Different versions to address the participants during facilitation: whole-school professional development session, PLC meeting (i.e., grade-level teams), or one-on-one sessions with individual teachers
What leadership knowledge do mathematics coaches have and/or need for effectively supporting teacher learning?	Set of survey items that align with effective mathematics leadership practices: • Items to describe mathematical knowledge needed for teacher leadership (Bitto, 2015) • Items to describe mathematical knowledge needed to support teacher development • Items to describe coaches' beliefs about teachers' mathematical knowledge for teaching • Observation instrument to capture evidence of the leadership mechanisms enacted by the mathematics coach while supporting teachers: • Protocol to describe mathematics coaching practices, questioning, and discourse
Does the MKT provide valid results for mathematics coaches?	Unknown if new instrument needed; a validation study on using the MKT with the mathematics coaching population is needed.

Note: Sample research questions and instruments provided. Additional options in text.

Yopp et al., 2008); however, instruments specifically designed to measure the leadership knowledge and skills that are essential for successful mathematics coaching are lacking. While measures of teachers' mathematical knowledge exist (e.g., Mathematical Knowledge for Teaching, Hill, Schilling, & Ball, 2004), these can be insufficient for describing the leadership knowledge exhibited and needed by coaches (Bitto, 2015).

As mathematics coaches and specialists are often viewed as teacher leaders (AMTE, 2013; Campbell & Malkus, 2013), assessments to address the leadership component of coaching, such as promoting professional learning, using data to drive instructional decisions, and leveraging research to improve instruction and learning (Teacher Leadership Exploratory Consortium, 2012) are necessary. While some general leadership frameworks and instruments are available (e.g., DuFour, 2004; Loucks-Horsey et al., 2010), the lack of connection to mathematics content and specific focus on coaching is problematic. A possible option to design a coaching leadership instrument would be the revision of an existing instrument coupled with the validation process with this new population.

Within the teacher-leader role, mathematics coaches must create an open, safe, and nonevaluative relationship between themselves and classroom teachers to: (1) provide feedback to teachers specific to mathematics teaching and learning, (2) develop a common vision and action plan for their school in regard to mathematics, and (3) mentor teachers to support high-quality mathematics instruction for all students (AMTE, 2013). Additionally, mathematics coaches are often charged with ensuring alignment between curriculum and assessments, evaluating equitable mathematics opportunities for all children, and working to reduce the learning discrepancies among student populations (AMTE, 2013). The infusion of mathematics content and practices, along with leadership knowledge and skills and are critical for exploring the relationships between these constructs (Bitto, 2015).

Instruments to measure this dynamic connection of leadership with a mathematics focus are needed to answer critical questions, including (1) What skills and practices better support effective mathematics coaching? (2) What specific leadership skills support effective coaching in a mathematics education context? and (3) How are we leveraging mathematics coaches' leadership knowledge and skills to impact change at the classroom and school levels? Instruments could include surveys of teachers about their interactions with the coach used alongside instruments for coaches to report on their work with teachers. For example, understanding which teachers the coach worked with the most (via a survey of the coach and teachers) and in what ways might inform decisions about how to best leverage coaches' knowledge to impact change in schools.

Instruments to Measure Communication and Facilitation

Although communication and facilitation play large roles in the leadership aspect of mathematics coaching, we believe that they should be highlighted as separate constructs of instrument development need as they are critical for the success of mathematics coaches and specialists. Mathematics coaches and specialists are often asked to serve as content facilitators to plan, deliver, and evaluate professional development, and subsequently to support the translation of teachers' new knowledge into their instructional practices (AMTE, 2013; Elliott et al., 2009; Lesseig et al., 2016). Mathematics coaches must engage in meaningful and purposeful dialogue to support teachers in the selection and use of materials, curriculum, and instructional practices. Additionally, mathematics coaches play a key role in the reflective practices of teachers by supporting, facilitating, and guiding teachers' reflections on specific lessons, practices, and assessment data. Effectively communicating to teachers, parents, students, and other stakeholders is a critical skill for mathematics coaches (AMTE, 2013).

There is a growing number of instruments to capture teacher practices in the mathematics classroom (e.g., Bostic et al., 2019; Boston & Wolf, 2006; Gleason et al., 2017), but the field is more limited on instruments to measure the practices of the coach as a facilitator during professional development at all levels (see Figure 8.2). Instruments are needed to measure: (1) interactions with teachers, including the dynamics involved in coaching conversations and coaching cycles; (2) facilitation of teachers' content and pedagogical content knowledge; and (3) leadership and facilitation of teachers' practices learned during professional development. These needed instruments must focus on connecting these facilitation and communication skills to the mathematics standards and practices to effectively measure their impact on the teaching and learning of mathematics. Such instruments could take the form of logs completed by both teachers and coaches participating in these interactions, as well as observation tools for capturing the interactions involved during coaching conversations or professional development session leadership (similar to classroom observation tools). While logs are self-report data, the work of coaching is difficult to observe as it may take place over entire days or weeks.

In addition to conversations and facilitation centered on mathematics, an emphasis on facilitating equitable mathematics practices and teaching should be an integral part of the mathematics coaches' role. Mathematics coaches should lead in advocating for all students, through their communication and facilitation of curriculum selection and alignment, teaching practices, and development of appropriate interventions, as needed (AMTE, 2013). Although the area of equitable mathematics practices is an understudied area in the facilitation work of mathematics coaches, instruments to support this work are becoming more readily available (e.g., the Equity QUantified In Participation [EQUIP] observation tool; Reinholz & Shah, 2018) and may provide an avenue to better support this important work of mathematics coaches. Although instruments such as EQUIP (Reinholz & Shah, 2018) were designed as a classroom observation tool, their potential to support coaches in their work is quite valuable and could potentially lead to the development of similar measures specifically designed to capture the facilitation of equitable instructional practices on a coaching level.

Instruments to Measure Roles, Responsibilities, and Impact

Mathematics coaches serve in a variety of roles in schools, as well as have a robust set of responsibilities. These roles and responsibilities may vary based on the state, district, and/or school (Campbell & Griffin, 2017; Fennell, 2017; McGatha & Rigelman, 2017b). Because of this shifting in roles and responsibilities, it is often difficult to capture what exactly mathematics coaches are doing in their daily work in schools. The inability to accurately capture mathematics coaches' roles and responsibilities also makes it difficult to directly link their work to changes in teacher practice and ultimately changes in student learning.

Instruments that proportionally measure and distinguish between the various roles and responsibilities are critical to establishing how mathematics coaches' work impacts teachers and students. Without being able to efficiently and accurately measure the roles and responsibilities mathematics coaches engage in, we are unable to determine what causal links to teacher practice and improved student performance exist. To date, some researchers have gathered information on the roles and responsibilities of mathematics coaches (e.g., Campbell & Griffin, 2017; Campbell & Malkus, 2010; Cobb et al., 2011; ems&tl, 2012); however, the vast differences in how and when this data are collected leave us with additional questions: (1) How do we define the roles and responsibilities of mathematics coaches in order to collect time spent engaged in this activities? (2) How often should this information be collected to adequately represent the various roles and responsibilities of mathematics coaches? (3) What is the most efficient way to collect this important data? Daily, weekly, monthly, yearly? and (4) What type of instrument (survey, interview, log, etc.) is most reliable and produces valid outcomes and interpretations for this type of data collection? Answering such questions is important for understanding how best to ultimately support student learning of mathematics via the support mathematics coaches provide to teachers.

As instruments are developed to capture the time mathematics coaches spend engaged in various roles and responsibilities, the question remains of whether one instrument can capture this for the varying titles and roles (see Figure 8.1) of mathematics coaches. For instance, the varying of roles and or job titles may necessitate different measures depending on the position, defined roles, context, and other factors. As with all of the instruments needed in this area of research, careful consideration on the intended audience is paramount. However, without a move toward developing instruments that produce valid and reliable interpretations and outcomes, the ultimate goal of improving students' mathematical learning through the support of mathematics coaches may be elusive. The needed measures discussed in this chapter will allow empirical research to continue to support the work of determining the most effective practices of mathematics coaches, the knowledge needed to be effective, and how mathematics coaches can support teacher practices and student learning and performance. With the use of these additional measures, we can work to substantiate the positive findings from previous work, expand on the current literature base, and provide evidence to support the use, training, and work of mathematics coaches.

Measurement Challenges and Future Considerations

As we consider research regarding mathematics coaches, there are many challenges for the development of instruments that produce valid interpretations and outcomes that fall into two major categories: (1) defining coaching roles and practices and (2) situating practice in context. The challenge to define coaching includes responding to the varied ways coaches work in schools and districts. These are subject to different policies between states/districts and the grassroots nature of coaching work where the coach, by design, has varied responsibilities in order to be responsive to current needs. Along with the challenges of definition and context, two broad areas of coaching research and instrument development are presented for future consideration. Specifically, we expand the discussion of the work on elementary mathematics coaches to include research and measurement considerations associated with mathematics coaches' identity and beliefs and the work of secondary mathematics coaches.

Challenge 1: Defining Roles and Practices

A challenge in measuring mathematics coaches' work is to determine what their primary role (or collection of roles) and work is within their school setting. Because of the terminology differences, it may not be enough to ask "Does your school have a mathematics coach?" as a survey item without knowing what the common title (or titles) may be in the local context. In addition, principals or district-level administrators may have a different understanding of what the mathematics coach does than what the mathematics coach understands about their work. Some states have certification and others do not; therefore, cross-state comparisons need to consider how the mathematics coach was hired or selected for the role they are taking on when making comparisons and gathering data.

Once the role itself is defined, another challenge involves mathematics coaches' leadership practices, including the complexity of connecting the coaching/leadership practices within a system of professional development (teacher outcomes, student outcomes, and/or school outcomes). Specifically, defining the role of the mathematics coach on a professional development project is a challenge. For example, are the teachers receiving a professional development intervention that is nested within one group that is led by one mathematics coach? Does one mathematics coach lead multiple groups of teachers within the same school or across different schools? How was the mathematics coach identified (e.g., if the state has certification, what was part of the decision-making)? An aspect of the leadership practice involves the perception of the coach as an administrator. While coaches may be involved with administrative tasks and support data analyses, it is cautioned that a mathematics coach should not generally take on an evaluative role (AMTE, 2013). A final challenge involves time. The measurement of time includes identifying categories that are longitudinal, appropriate time scales for measurement and score interpretation. For example, suppose the researcher would like to understand how the coach can best support grade-level teams. Documenting such work could require long-term, multiyear collection of evidence of practice and data from multiple sources (e.g., coach, teachers, principal). Instruments should account for repeated measurement and change over time.

Challenge 2: Situating Coaching Work in Context and Purpose

When designing measures to capture the complex nature of the work mathematics coaches, we offer a few specific ideas to keep in mind for situating the coaching work of interest in context. Returning to Figures 8.1 and 8.2, the project needs to define the model being used for the relationship between the mathematics coach and teachers in the professional development experience. These measurement challenges would shift depending on the role and responsibility of the coach. For instance, in a coach–student relationship, the measurement challenges and considerations would be quite different from the ones posed in this chapter, where we discuss the coach–teacher relationship. Defining this relationship is important for understanding if data are needed from teachers to align with data from coaches. A consideration for measurement design and data collection is how the data from teachers and coaches are collected and analyzed.

Three important considerations are presented here and are dependent on the coach relationship. First, having a clear definition of how specialists or coaches are being used in your specific context and sharing this definition in your work is crucial. Without this, we are left wondering, who is this measure valid for? For what context was this instrument created? Second, having a clear purpose for the intended use of the instrument is essential as well. Is the instrument designed to measure constructs related to the knowledge and skills of mathematics coaches? Is the instrument designed to support the work of coaches with teachers, groups of teachers, or at the school level? How, when, and by whom should the instrument be used for data collection and analysis? How should scores be interpreted? Third, it is also likely that measures of mathematics coaching would be used with measures of student/teacher learning and development, so the connections across measures need to be considered. For validity and reliability, this could mean asking how questions on a teacher measure align with questions on a mathematics coach survey.

Future Considerations for Mathematics Coaching Research

Identify and Beliefs

An area for which we have not identified instruments for mathematics coaching is measures of their identities and beliefs related to coaching. As with other instruments, there may be measures of teachers' beliefs about mathematics teaching and learning, but we need to examine those beliefs from a teacher leadership perspective as well. In terms of identity, coaches go through a shift from being a teacher of children to working with adults (Chval et al., 2010). In conceptualizing this challenge, Chval et al. describe their role as supporting teachers, students, and the school at large, which shifts coaches from being experienced teachers to novices as mathematics coaches. As they stated, "Too often we assume that effective teachers will be effective coaches and that these teachers need little support as they transition in their new roles as mathematics coaches." (p. 192, Chval et al., 2010). Campbell and Griffin (2017) point to the need for research about the change in identity when shifting from teacher in a school to being a coach in the same school. Instruments are needed to describe these transitions, to document the knowledge needed to shift from expert teacher to coach, and to evaluate the supports that effectively support the transition.

Secondary Mathematics Coaches

Although much of the work discussed focuses on elementary mathematics coaches, some of the same measurement issues arise when targeting the work of mathematics coaches at the middle and/or high school levels. Secondary mathematics coaches have received far less attention than their elementary counterparts; however, an ongoing need to support teachers at this level is clear (Wray & Barnes, 2017). Research on secondary mathematics coaching not only needs instrumentation to address the areas discussed in this chapter, but also to validate the measures that are currently available and being used at the elementary level. Because much of the work of secondary mathematics coaches builds from the work of elementary mathematics coaches (Wray & Barnes, 2017), instrumentation may be able to cross grade-band lines to provide valid interpretations and outcomes on mathematics coaching. Research is needed to address this possibility and the concerns.

Conclusion

The chapter has explored some of the validity challenges inherent in documenting and studying the complexity of mathematics coaching practice and impact. Coaching itself is still an emerging role in schools and schools are still experimenting with how best to incorporate these teacher leaders in school-based professional development. As a result, coaching research is building a knowledge base about the possible impact of coaching while also building tools to document that impact. It is both a benefit of coaching that coaches are able to respond to school-based needs, but also a challenge since practices and roles may not be clear. The collection of evidence about the roles, responsibilities, and impact of mathematics coaches has the potential to inform state and district policies about licensure, resource allocation, and job roles and responsibilities.

As we consider instrument development and measures currently being used, we must acknowledge that instruments are designed with a particular purpose and intended audience, although are often being used for a different purpose or with a different intended audience within mathematics coaching research. This poses great concerns when thinking about the validity of the outcomes and interpretations of this work. There is currently a lack of recognition that the work of mathematics coaching is different from the work of teaching children. However, this recognition of the unique work of mathematics coaches is critical to advance the field. Therefore, the instruments developed for the use of research with mathematics coaches may need to be entirely different, interpreted differently, or validated with the mathematics coach population. The validity question for the design and use of mathematics coaching instruments is whether they are measuring the constructs as intended and whether the subsequent analysis captures the complex work of coaching. When there is a lack of measures to use, researchers may attempt to retrofit existing instruments or create their own without attending to a validation process; therefore, these instruments being used are likely not producing evidence and results that are valid and reliable. A second layer of validity is addressing how the relationship between teachers and coaches is captured to understand instructional change and, in turn, student learning. The field is in need of large-scale, multistate, multidistrict comparisons in order to add to the qualitative research regarding the impact of mathematics coaches.

As we move forward in the use and research of mathematics coaches, we bring this call to action to the mathematics education community: We must devote time to considerations of measurement in the coaching context. The need for developing and using measures that produce valid and reliable interpretations and outcomes is fundamental in the future study of mathematics coaches. Critical work is already underway to provide evidence of the use of mathematics coaches as a means to improve the teaching and learning of mathematics (see McGatha et al., 2017 for an overview of research); but now, we must come together in an effort to move the field forward with thinking through validity concerns, constructing measures to capture various aspects of their work, and aiming to advance the complex and nuanced work of mathematics coaches.

Note

1. The standards documents use the term *specialist,* but the nature of the work described is consistent with our definition of *coach.*

References

Association of Mathematics Teacher Educators. (2013). *Standards for elementary mathematics specialists: A reference for teacher credentialing and degree programs.* San Diego, CA: AMTE.

Baldinger, E., & Louie, N. (2014). *TRU Math conversation guide: A tool for teacher learning and growth.* Berkeley: Graduate School of Education, University of California; Berkeley/East Lansing, MI: College of Education; Michigan State University. Retrieved from http://ats.berkeley.edu/tools.html and/or http://map.mathshell .org/materials/pd.php

Barlow, A. T., Burroughs, E. A., Harmon, S. E., Sutton, J. T., & Yopp, D. A. (2014). Assessing views of coaching via a video-based tool. *ZDM, 46,* 227–238. https://doi.org/10.1007/s11858-013-0558-7.

Berry, R. Q., Rimm-Kaufman, S. E., Ottmar, E. R., Walkowiak, T. A., & Merritt, E. G. (2010). *The mathematics-scan coding guide.* Unpublished measure, University of Virginia.

Bitto, L. E. (2015). *Roles, responsibilities, and background experiences of elementary mathematics specialists* (Doctoral dissertation). Retrieved from ProQuest Dissertations & Theses Global. (1687831827).

Borko, H., Jacobs, J., Koellner, K., & Swackhammer, L. E. (2015). *Mathematics professional development: Improving teaching using the problem-solving cycle and leadership preparation models.* New York, NY: Teachers College Press.

Borko, H., Koellner, K., & Jacobs, J. (2014). Examining novice teacher leaders' facilitation of mathematics professional development. *The Journal of Mathematical Behavior, 33,* 149–167. http://dx.doi.org/10.1016/j.jmathb.2013.11.003

Bostic, J., Lesseig, K., Sherman, M., & Boston, M. (in press). Classroom observation and mathematics education research. *Journal of Mathematics Teacher Education.*

Bostic, J. D., Matney, G. T., & Sondergeld, T. A. (2019). A validation process for observation protocols: Using the *Revised SMPs Look-for Protocol* as a lens on teachers' promotion of the standards. *Investigations in Mathematics Learning, 11*(1), 69–82. doi:10.1080/19477503.2017.1379894

Boston, M., Bostic, J., Lesseig, K., & Sherman, M. (2015). A comparison of mathematics classroom observation protocols. *Mathematics Teacher Educator, 3,* 154–175.

Boston, M., & Wolf, M. K. (2006). *Assessing academic rigor in mathematics instruction: The development of the instructional quality assessment toolkit: CSE Technical Report 672.* Los Angeles: University of California, National Center for Research on Evaluation, Standards, and Student Testing (CRESST).

Campbell, P. F., Ellington, A. J., Haver, W., & Inge, V. (2013). *Elementary mathematics specialist's handbook.* Reston, VA: National Council of Teachers of Mathematics.

Campbell, P. F., & Griffin, M. J. (2017). Reflections on the promise and complexity of mathematics coaching. *Journal of Mathematical Behavior, 46,* 163–176. https://doi.org/10.1016/j.jmathb.2016.12.007

Campbell, P. F., & Malkus, N. N. (2011). The impact of elementary mathematics coaches on student achievement. *The Elementary School Journal, 111,* 430–454.

Campbell, P. F., & Malkus, N. N. (2013). The mathematical knowledge and beliefs of elementary mathematics specialist-coaches. *ZDM, 46,* 213–225. https://doi.org/10.1007/s11858-013-0559-6

Center for Research in Mathematics and Science Teacher Development (CRMSTD) University of Louisville. (2008). *Diagnostic mathematics assessments for elementary school teachers.* Retrieved from http://louisville.edu/education/centers/crimsted/diag-math-assess-elem

Chval, K. B., Arbaugh, F., Lannin, J. K., van Garderen, D., Cummings, L., Estapa, A. T., & Huey, M. E. (2010). The transition from experienced teacher to mathematics coach: Establishing a new identity. *Elementary School Journal, 111,* 191–216.

Cobb, P. A., Henrick, E. C., & Munter, C. (2011, April). *Conducting design research at the district level.* Paper presented at the annual meeting of the American Educational Research Association, New Orleans, LA.

Cobb, P. A., & Jackson, K. (2011). Towards an empirically grounded theory of action for improving the quality of mathematics teaching at scale. *Mathematics Teacher Education and Development, 13*(1), 6–33.

Cobb, P. A., & Jackson, K. (2015). Supporting teachers' use of research-based instructional sequences. *ZDM, 47,* 1027–1038. https://doi.org/10.1007/s11858-015-0692-5

Collins, A., Brown, J. S., & Newman, S. E. (1987). *Cognitive apprenticeship: Teaching the craft of reading, writing, and mathematics* (National Institute of Education Technical Report No. 403). Urbana, IL: Center for the Study of Reading.

Costa, A. L., & Garmston, R. J. (1994). *Cognitive coaching: A foundation for renaissance schools.* 480 Washington, Street, Norwood, MA: Christopher-Gordon Publishers, Inc.

Darling-Hammond, L., Hyler, M. E., & Gardner, M. (2017). *Effective teacher professional development.* Palo Alto, CA: Learning Policy Institute.

Dean, C., & McClain, K. (2006, April). *Situating the emergence of a professional teaching community within the institutional context.* Paper presented at the annual meeting of the American Educational Research Association, Chicago, IL.

Dossey, J. A. (1984). One point of view: Elementary school mathematics specialists: Where are they? *Arithmetic Teacher, 32*(3), 3–50.

DuFour, R. (2004). Leading edge: Leadership is an affair of the heart. *Journal of Staff Development, 25*(1), 1–5.

Elementary Mathematics Specialists and Teacher Leaders Project (ems&tl). (2012). *Annual project evaluative report.* Waukesha, WI: The Brookhill Foundation.

Elliott, R., Kazemi, E., Lesseig, K., Mumme, J., Carroll, C., & Kelley-Petersen, M. (2009). Conceptualizing the work of leading mathematical tasks in professional development. *Journal of Teacher Education, 60,* 364–379. https://doi.org/10.1177/0022487109341150

Felux, C., & Snowdy, P. (2006). *The math coach field guide: Charting your course* (1st ed.). Sausalito, CA: Math Solutions.

Fennell, F. (2017). We need mathematics specialists now: A historical perspective and next steps. In M. B. McGatha & N. R. Rigelman (Eds.), *Elementary mathematics specialists: Developing, refining, and examining programs that support mathematics teaching and learning* (pp. 3–18). Charlotte, NC: Information Age Publishing, Inc.

Fennell, F. (Skip), Kobett, B. M., & Wray, J. A. (2017). Elementary mathematics specialists and teacher leader project. In M. B. McGatha & N. R. Rigelman (Eds.), *Elementary mathematics specialists: Developing, refining, and examining programs that support mathematics teaching and learning* (pp. 115–122). Charlotte, NC: Information Age Publishing, Inc.

Fullan, M. S., & Stiegelbauer, S. S. (1991). *The new meaning of educational change.* London, England: Cassell.

Gibbons, L. K. (2017). *Examining Mathematics Coaching: Practices that help develop school-wide professional learning.* Charlotte, NC: Association of Mathematics Teacher Educators.

Gleason, J., Livers, S. D., & Zelkowski, J. (2017). Mathematics classroom observation protocol for practices (MCOP2): A validation study. *RCML Investigations in Mathematics Learning, 9,* 111–129. https://doi.org/10.1080/1947750 3.2017.1308697

Greeno, J. G., Moore, J. L., & Smith, D. R. (1993). Transfer of situated learning. In D. K. Detterman & R. J. Sternberg (Eds.), *Transfer on trial: Intelligence, cognition, and instruction* (pp. 99–167). Westport, CT, US: Ablex Publishing.

Guskey, T. R. (2000). *Evaluating professional development.* Thousand Oaks, CA: Corwin Press Inc.

Hall, G. E., Wallace, R. C., & Dossett, W. F. (1973). *Procedures for adopting educational innovations.* Austin, TX: The Research and Development Center for Teacher Education.

Harbour, K. E., Adelson, J. L., Karp, K. S., Pittard, C. M. (2018). Examining the relationships among mathematics coaches and specialists, student achievement, and disability status: A multi-level analysis using NAEP data. *The Elementary School Journal, 188,* 654–679. https://doi.org/10.1086/697529

Hill, H. C. (2010). Mathematical Quality of Instruction (MQI). *Learning Mathematics for Teaching*. University of Michigan, unpublished manuscript.

Hill, H. C., Schilling, S. G., & Ball, D. L. (2004). Developing measures of teachers' mathematics knowledge for teaching. *Elementary School Journal, 105*, 11–30.

Hill, H. C., Sleep, L., Lewis, J., & Ball, D. (2007). Assessing teachers' mathematical knowledge: What knowledge matters and what evidence counts. In F. Lester, Jr. (Ed.), *Second handbook of research on mathematics teaching and learning: A project of the National Council of Teachers of Mathematics* (pp. 111–156). Charlotte, NC: Information Age Publishing.

Hipp, K. K., & Huffman, J. B. (2003, January). *Professional learning communities: Assessment-development-effects*. Paper presented at the International Congress for School Effectiveness and Improvement, Sydney, Australia.

Hopkins, M., Ozimek, D., & Sweet, T. M. (2017). Mathematics coaching and instructional reform: Individual and collective change. *The Journal of Mathematical Behavior, 46*, 215–230. https://doi.org/10.1016/j.jmathb.2016.11.003

Joyce, B., & Showers, B. (2002). *Student achievement through staff development* (3rd ed.). Alexandria, VA: ASCD.

Kanold, T. D., & Larson, M. R. (2012). *Common core mathematics in a PLC at works: Leader's guide*. Bloomington, IN: Solution Tree Press.

Killion, J., Harrison, C., Bryan, C., & Clifton, H. (2012). *Coaching matters*. Oxford, OH: Learningforward.

Lave, J., & Wenger, E. (1991). *Situated learning: Legitimate peripheral participation*. Cambridge, England: Cambridge University Press.

Lesseig, K., Elliott, R., Kazemi, E., Kelley-Petersen, M., Campbell, M., Mumme, J., & Carroll, C. (2016). Leader noticing of facilitation in videocases of mathematics professional development. *Journal of Mathematics Teacher Education, 20*, 591–619. https://doi.org/10.1007/s10857-016-9346-y

Loucks-Horsley, S., Stiles, K. E., Mundry, S., Love, N., & Hewson, P. W. (2010). *Designing professional development for teachers of science and mathematics*. Thousand Oaks, CA: Corwin.

Markworth, K. A., Brobst, J., Ohana, C., & Parker, R. (2016). Elementary content specialization: Models, affordances, and constraints. *International Journal of STEM Education, 3*(1), 16. https://doi.org/10.1186/s40594-016-0049-9

McGatha, M. B., & Bay-Williams, J. M. (2013). Making shifts toward proficiency. *Teaching Children Mathematics, 20*, 162–170. doi:10.5951/teacchilmath.20.3.0162

McGatha, M. B., Davis, R., & Stokes-Levine, A. (2017). Mathematics specialists: What does the research say? In M. B. McGatha & N. R. Rigelman (Eds.), *Elementary mathematics specialists: Developing, refining, and examining programs that support mathematics teaching and learning* (pp. 91–100). Charlotte, NC: Information Age Publishing, Inc.

McGatha, M. B., & Rigelman, N. R. (2017a). *Elementary mathematics specialists: Developing, refining, and examining programs that support mathematics teaching and learning* (Vol. 2). Charlotte, NC: Information Age Publishing.

McGatha, M. B., & Rigelman, N. R. (2017b). Introduction. In M. McGatha & N. R. Rigelman (Eds.), *Elementary mathematics specialists: Developing, refining, and examining programs that support mathematics teaching and learning* (pp. xiii–xv). Charlotte, NC: Information Age Publishing.

Munter, C. (2014). Developing visions of high-quality mathematics instruction. *Journal for Research in Mathematics Education, 45*, 584–635. https://doi.org/10.5951/jresematheduc.45.5.0584

National Assessment for Educational Progress. (2019). Reading, mathematics, and science school questionnaire. *2019 operational school characteristics and policies grade 4 school questionnaire.* National Center of Education Statistics, Institute of Education Sciences. Retrieved from https://nces.ed.gov/nationsreportcard/experience/survey_questionnaires.aspx

National Council for Teachers of Mathematics. (2012). *NCTM CAEP standards (2012): Elementary mathematics specialist (advanced preparation).* Retrieved from www.nctm.org/Standards-and-Positions/CAEP-Standards/

National Council of Supervisors of Mathematics. (2008). *The PRIME leadership framework: Principles and indicators for mathematics education leaders.* Bloomington, IN: Solution Tree.

Reinholz, D. L., & Shah, N. (2018). Equity analytics: A methodological approach for quantifying participation patterns in mathematics classroom discourse. *Journal for Research in Mathematics Education, 49*, 140–177. doi:10.5951/jresematheduc.49.2.0140

Schoenfeld, A. H., Floden, R. E., & The Algebra Teaching Study and Mathematics Assessment Project. (2014). *An introduction to the TRU Math Dimensions.* Berkeley, CA & E. Lansing, MI: Graduate School of Education, University of California, Berkeley & College of Education, Michigan State University. Retrieved from http://ats.berkeley.edu/tools.html and/or http://map.mathshell.org/materials/pd.php

Secada, W., & Byrd, L. (1993). *Classroom observation scales: School level reform in the teaching of mathematics.* Madison, WI: National Center for Research in Mathematical Sciences Education, Wisconsin Center for Education Research, School of Education.

Smith, M. (2001). *Practice-based professional development for teachers of mathematics.* Reston, VA: National Council of Teachers of Mathematics.

Sowder, J. T. (2007). The mathematical education and development of teachers. In F. Lester (Ed.), *Second handbook of research on mathematics teaching and learning* (pp. 157–224). Reston, VA: National Council of Teachers of Mathematics.

Teacher Leadership Exploratory Consortium. (2012). *Teacher leader model standards.* Retrieved from www.teacherleaderstandards.org/standards_overview

Wainwright, C. L., Flick, L. B., & Morrell, P. D. (2003). Development of instruments for assessment of instructional practices in standards-based teaching. *Journal of Mathematics and Science: Collaborative Explorations, 6*(1), 21–46.

West, L., & Cameron, A. (2013). *Agents of change: How content coaching transforms teaching and learning.* Portsmouth, NH: Heinemann.

West, L., & Staub, F. C. (2003). *Content-focused coaching: Transforming mathematics lessons.* Portsmouth, NH: Heinemann.

Whitenack, J. W., & Ellington, A. J. (2007, April). *A methodology to explain teachers' emerging roles as K-5 mathematics specialists.* Paper presented at the annual meeting of the American Educational Research Association Conference, Chicago, IL.

Woulfin, S. L., & Rigby, J. G. (2017). Coaching for coherence: How instructional coaches lead change in the evaluation era. *Educational Researcher, 46*, 323–328. https://doi.org/10.3102/0013189X17725525

Wray, J. A., & Barnes, W. J. (2017). A case for secondary mathematics coaches. In M. B. McGatha & N. R. Rigelman (Eds.), *Elementary mathematics specialists: Developing, refining, and examining programs that support mathematics teaching and learning* (pp. 91–100). Charlotte, NC: Information Age Publishing, Inc.

Yopp, D. (2008a). The Examining Mathematics Coaching project, Coaching Impact Instrument (CII), unpublished manuscript.

Yopp, D. (2008b). The Examining Mathematics Coaching project, Coaching Skills Inventory (CSI), unpublished manuscript.

Yopp, D., Rose, J., & Meade, C. (2008). The Examining Mathematics Coaching project, Coaching and Teacher Reflection Instrument (CTRI), unpublished manuscript.

Index

assessment 2

BEAR Assessment System 65–71, 84
bias 25, 81–84

classroom assessment 14
Common Core State Standards 65,
    147
Comprehensive Assessment of
    Outcomes 160
consequences from testing 30, 42,
    80–81, **128**, 141, **162**, 164
consequential validity evidence 159
construct 124
construct map 17–19, 51–52, 66

ecological validity evidence 47
elemental validity evidence 53
elementary 50, 173, 178–179
external validity evidence 158–159

factor 124–125
fairness *see* bias

generalizability study 106–109
generalizability validity evidence
    157–158
Guidelines for Assessment and
    Instruction in Statistics Education
    154–155

inservice teacher 14, 42–43, 92–93,
    160, 172–173
instructional log 179
Instructional Practice Log in
    Mathematics 94–95
internal structure 2, 29, 42, 75, **128**,
    136–140, **162**, 163–164

interpretive use argument 91,
    114–115
interview 180

LOCUS 140, 159, 161

mathematics coach 172, *174*, *175*,
    181, 186–189
Mathematics SCAN 93–95
Mathematics Teaching Efficacy Beliefs
    Instrument 15
measure 2
Messick, S. J. 90–91, 147–148, 152
middle grades 158

observation protocol 15, **32**, 34, 93,
    178, 184

pilot testing 33
preservice teacher 158
purpose statement 3, 16–17

Rasch modeling 74–75
rater coding 34, 103
relations to other variables 29, 42,
    79–80, **128**, 140–141, **162**, 164
reliability 30, 137
response processes 28–29, 42, 74–75,
    128, 140–141, **162**, 164
Revised Standards for Mathematical
    Practice Look-for Protocol 15, 20

secondary 158, 188–189
self-efficacy 127
Standards for Educational and
    Psychological Testing 1, 3, 12, 42,
    64, 73, 84, 90–91, 126, 163–164
statistical thinking 149

statistics teaching efficacy 149
structural validity evidence 47
substantive validity evidence
    156–157
summary statement *see* purpose
    statement
survey 179–181

teacher beliefs 148
teacher education 160
teacher self-efficacy 148
test *see* measure

test content 2, 28, 42, 72–74, **128**–133,
    154–156, **162**, 163

undergraduate student 121

validity 2, 41–42, 64, 84, 90, 126, 152
Validity and Measurement in
    Mathematics Education 6, 16, 85
validity argument framework **73**, **96**

Wilson, M. 65
Wright map *68–71*, 134, 157